To
Kyl
Keep'em
Flying
1/24/2015

The Fun of Flying

The Fun of Flying

Frits Forrer

Holland's Glory P.O. Box 488
Gulf Breeze, Florida 32562

Contact author at: fritsforrer@cox.net
Phone 850-384-5363
Home 850-607-414
www.fritsforrer.net

This edition was prepared for printing by
Ghost River Images
5350 East Fourth Street
Tucson, Arizona 85711
www.ghostriverimages.com

Cover design by: Betty Shoopman
bcshoop@bellsouth.net

ISBN-10 0-9714490-3-1
ISBN-13 978-0-9714490-3-9

Library Of Congress Control Number: 2412750338

Printed in the United States of America
Revised January, 2012

DEDICATED

To all those who were able, dedicated and excited about FLYING, but paid the ultimate price, their lives, while doing what they liked best:

BEING UP THERE IN THEIR FLYING MACHINES

Much gratitude to those who helped critique this book that was originally written *with a Dutch accent.*

Phil Martin, very patient English Language Expert Extraordinaire!

Art Giberson, author of many Navy books, Documentaries and Novels.

Rien Kamphuis, General (Commodore) of the Royal Netherlands Air Force.

Katjang Wildbergh, fellow F-84 pilot and old drinking buddy.

Massey Lambard, USAF F-84 pilot and author.

Katy, my wife, whose spelling ability is beyond the norm.

Edye Brewis, my most enthusiastic fan and editor.

Betty Shoopman, graphic artist and friend, who designed the cover.

They all contributed to make the book exciting and *readable!*

Thanks,

Frits

Other books by Frits Forrer

Five Years Under The Swastika
ISBN 0-9714490-0-7

Someone's Smiling Down On Us
ISBN 0-9714490-1-5

Tampa Justice, No Money, No Justice
ISBN 0-9714490-5-8

Smack Between The Eyes
ISBN 0-9714490-6-6

To Judge or Not To Judge
ISBN 0-9714490-7-4

The Golden Pig
ISBN 0-9714490-8-2

Brothers Forever
ISBN 0-9714490-9-0

The Curse Of The Black Mamba
ISBN 978-0-9822207-0-2

Golden Nuggets
ISBN 978-0-9822207-1-9

Thunderclouds Over America
ISBN 978-0-9822207-2-6

Suffering With A Smile
ISBN 978-0-9822207-3-3

Forty Million
ISBN 978-0-9822207-4-0

www.fritsforrer.net

Contents

CHAPTER I
THE DRAFT

"Drafted?" "Not Me!"

The line of 18 year-olds was moving slowly toward the cubby-holes in the next room.

"Why shouldn't you get drafted? What makes you so exceptional?"

Two boys, former classmates, have met up again after three years of not seeing each other.

"Well, for one, I have flat feet, so I won't even pass this physical. Besides that, my older brother had his physical six months ago, so he'll be drafted and I'll be exempt because of him, and third, my dad is sick - my Mam needs my income and help in the business and they have some sort of excuse for that."

"Well, I'll help you hope so. I may not be as lucky as you are, they'll grab my ass for sure!"

The line moves up another notch.

"By the way Frits, what do you do for a living and fun?"

"I'm an office-machine apprentice mechanic in a union, so I don't make much money, but it's interesting!"

"Really? You were so good at math and geometry, I always thought you'd become an engineer."

"Hell man, I wanted to and I applied and was accepted, but we didn't have money and I needed to go to work and help support the family. So, I went to work. How about you, Bert?"

"I work with my Dad in our print shop in Bergen and I still go to school nights, learning graphics."

"Graphics?"

"Yeah, you know, designs, artwork and stuff like that. I like it. Oh wait, the doctor wants me, wait for me afterwards, we'll shoot some billiards. See you later."

"Take a deep breath. Hold it. Fine, breathe out. Good. Sit up now, swing your legs off the table. Good."

"Hey doctor, I have flat feet, so I'll be 4-F, right?"

"You have slightly fallen arches, son, but your feet are fine. ***Next please!?"***

"But Doc ... "

"Next please."

"Damn Doctor!"

"Sit over here please."

"Yes, Captain."

"I'm a Corporal, Mr. Forrer. Just checking your application."

"I want you to know Corporal, I'm not applying and I won't be drafted!"

"That's fine, Mr. Forrer. Is your address correct here? St. Pancras? You want to be a Navy Pilot or an Air Force Pilot?"

"Yep!"

"There are two more choices to fill in."

"O.K., put down Air Force Pilot and Navy Pilot. The Army and Marines don't have pilots, do they? That's what I thought. You want four spaces filled in. Fill'em in."

"Sports Mr. Forrer? Soccer, right? Swimming, right? Bike racing, right? Schooling? You have a M.U.L.O. diploma, A & B, right? O.K., you'll be hearing from us."

"There's NO way, I'll ever be in the service. I'm a hardship case and I have a brother exemption. But it was nice of you to invite me Corporal. Have a nice day."

Two months later, October 1, 1950, Frits' brother Joop (pronounce yope) was drafted in the Army. He was madder than a pistol. He'd been for a Flight Physical, spent an additional two days in what he called PSYCHO-TECHNICAL TESTING, passed the physical but was turned down for flight training.

Now the Army had drafted him.

His girlfriend's brother, Karel, was flying "GLOSTER-METEORS, the first JET-FIGHTERS of the ROYAL NETHERLANDS AIR FORCE, and that's all Joop had wanted to do. When that didn't work, he applied for hardship exemption, with his dad being ill and his Mam needing Joop's income.

"Denied". Drafted In The Army!" "What a stinking, low-down, rotten deal!"

In November Frits received an invitation for a physical exam for pilot training at the selection center in SOESTERBERG. For four hours he blew up Mercury columns, blew up a tank with air, lined up sticks in the distance and jogged for blood pressure readings and pulse checks. He looked at little letters, large letters, colored letters, opaque letters and inverted letters. He gave blood, peed in a glass and went home, informed that he was a healthy specimen, ready for flight in any type of aircraft available.

A few weeks later he found himself in YPENBURG, analyzing inkblots, talking with psychologists, psychiatrists and other dumb people. He had to lead a team of "Buddy-Victims" across a pit, up a wall and then to safety, making like **Tarzan**. He did some algebra and geometry puzzles and got back on the train for home.

Later he explained to Joop; "During the two days that I spent there, they did not give me ONE question or assignment, that an eighth grader couldn't have handled and that's why I passed and you didn't." He smiled broadly.

"You see, you were given assignments and you attacked them as if they were highly intelligent problems. I looked at them as if they were really dumb questions, with simple solutions and I treated them accordingly. The fact that I have this piece of paper from the Air Force proves me right."

During Christmas leave from the Army, Joop told Frits a lot about military life and how he should approach it if he got drafted too. Not expecting to get drafted, he forgot half the advice.

When the notice and the train ticket arrived, he went out, bought a dozen locks, a gross of condoms and a 20 pack box of playing cards. That was part of the advice he remembered.

"BY ORDER OF HER MAJESTY THE QUEEN, THE NETHERLANDS AIR-FORCE REQUESTS THE HONOR OF YOUR PRESENCE AT NIJMEGEN TRAINING CAMP ON MARCH 10, 1951. YOU ARE TO LEAVE ON THE FIRST AVAILABLE PUBLIC TRANSPORTATION AFTER 8 AM."

It sounded good enough to have expected a limousine.

With mixed emotions, he rode the train. In a way it seemed exciting. At age 19, he was in somewhat of a rut. On the other hand, he cherished his freedom and that freedom was just being infringed upon.

Besides that, his Dad hated the military. He, himself, had been a draftee who served from 1914 till 1919, even though Holland was neutral during WORLD WAR 1. At one point, sitting atop a horse-drawn supply wagon, waiting for an inspection by Queen Wilhelmina, his feet were partially frozen. That made him hate the Queen and the military and it made his ulcers act up whenever his feet got cold.

The food, the pay, the living conditions and the frozen feet had made him most anti-military, to the point that while living in Belgium for eight years, he refused to become a Belgian citizen. Business-wise it would have been to his advantage, but Belgium had a two year draft at that time and Holland only a six-month's draft. He figured that NO WAY would his sons be subjected to two years in the service. That was before WORLD WAR II. Now, two of his three boys were already drafted for at least 18 months and any type of training would extend that to two years.

Non-Commisioned-Officers training, Officer's training, technical or specialist training would immediately tie somebody down for 24 months.

All he was thinking about, though, as the train rolled smoothly through the flat Dutch landscape was that Nijmegen was supposed to be a swinging city, more than 2,000 years old with a lot of history and culture, but also a bustling modern city with a lot of action. The "action" part was mostly on his mind, the historic part he'd seen a few times already.

By the time the train had made a number of stops the number of ***draftees*** versus other passengers increased steadily, and when the train finally pulled into the "rebuilt" railroad station in Nijmegen 800 young men piled out, looking like a wild painting.

All sizes and shapes, dressed in all sorts of clothes, they were caught by a group of blue-uniformed officers and non-coms as they filed out of the station.

Lined up in "THREE'S", they got marched off. Well, "MARCHED" was too strong a term. They shuffled off in the same general direction and after a twenty-minute march, they were inside the "GARRISON" compound.

Tables, lined up in the drill square, invited the new recruits to the "A-D, E-G, H-K, etc. sign of their choice and have their names matched with the ones on the list. A slip with a bunch of numbers steered them to the right BUILDING: "A", the right FLOOR: "2", the right ROOM: "2-1" and even the right BUNK: "13-lower."

Upon arrival in the room, the suitcases were dumped underneath the appropriate bunks and everyone was handed a blue checkered mattress cover and directions where to find the straw with which to fill the cover.

The Corporal, overseeing the ***"STUFFING"*** didn't agree with Joop's advice: "The more straw you stuff in it, the longer it takes to become as hard as a board."

"Just a little will do" was the Corporal's order, but Frits decided his brother's advice was probably more honest.

At ***"SUPPLY"*** they got to stand in line for their pillows, pillow cases, blankets and towels. Back in "2-1" a mimeographed sheet told them where to line up for haircuts and objects to purchase from the P.X., locks, toothbrushes, toothpaste, combs, etc.

Frits hollered across the room. "I have locks, CHEAP!" The dozen locks he brought went fast. He had only a 100% mark-up on them. Guys returning from the P.X. complained about long lines, so that helped his business. He regretted not having bought more than 12 locks, because of their weight.

Mess-tins were issued and filled with some sort of military gook, and immediately everybody did what every good ***draftee*** should do: BITCH, BITCH, BITCH. BITCH about the food, BITCH about the bed, the haircut, BITCH about everything, even when it's good. BITCH, BITCH, BITCH, even if it's better than anything you had at home: BITCH. Always BITCH. Well, you wouldn't really be a good serviceman if you didn't BITCH.

By mid-afternoon, Room "2-1", Building "A", was ready for something. Not war, not just yet, but at least 52 double bunks that lined the walls, were made with checkered mattresses, sheets and pillowcases. All new soldiers were in uniform and all lockers were neatly stacked with essentials. The rest was hidden away in the suitcases underneath the beds. Some of the guys didn't get their haircuts yet and some still needed locks, but otherwise the place looked like a military barracks.

A sergeant-first-class, an older fellow, hollered *"ATTENTION!"* a lot, while standing on a table. It took him a while to get everyone to stand in front of their own bunks, "LOWER" on the left, "UPPER" on the right.

A younger Sergeant, maybe 21, and a Corporal, another youngster, marched up and down the aisles created by the tables in the center and the bunks against the walls. They stopped by every recruit, buttoning a button, squaring a cap or angling the feet properly.

The older Sergeant, an obvious career man, hollered: *"SHUT-UP" and "QUIET"* a lot, interspersed with *"SILENCE!"* Finally he had their attention and ordered everyone to answer: *"PRESENT"* when their name was called.

Only five missing. Out of 104 that was not bad apparently, because he then hollered: (This man has one tone of voice, LOUD, Frits concluded.) *"When I say: "Parade Rest" you can spread your legs, fold your hands behind your back, but otherwise not move or talk!"*

"PARADE ... ... REST!"

Somewhat of a shambles occured but after a few more *"SHUT-UPS"* he regained control.

"I'm Sergeant Van Kluif, assisted by Sergeant Bongers and Corporal Mandhof. Your Squadron Commander is Lieutenant Zeltman and when he or I or any other officer or NCO appears, the first one to spot us is to holler: "ATTENTION" wherever you are and whatever you're wearing. Only one exception: If you're under the covers, STAY THERE!" Out of the corner of his mouth, Frits whispered to his bunkie: *"And lie at attention? Or stand your ding-a-ling at attention?"*

"Here now is Lt. Zeltman. ATTENTION!"

Lt. Zeltman replaced Sgt. Van Kluif on the table. *"When I say: "At Ease" you can relax, but no TALKING! AT EASE!"*

"I'm Lieutenant Zeltman, (obviously an Indonesian "RIJST-EPIKKER", Frits thought) *and you've met my staff. For the next three months, we're going to be together day and night, in order to make soldiers out of you. THIS MAY SURPRISE YOU!! WE SUCCEED AT THIS!!!!"*

"In every draft there are some short ones, some tall ones, fat ones and skinny ones, bright ones and dumb ones, cooperative ones and wise guys. At the end, all of you will be soldiers. The SMART ones, the cooperative ones will have a relatively good time at this. Some of the other ones will be punished unmercifully, but will make it anyway. THE INTERESTING THING IS ... IT IS YOUR CHOICE!"! The Lieutenant was hollering by then. *"You decide how pleasant or how miserable your stay with us is going to be. We don't care. We're here anyway.*

"You do have a disadvantage in this squadron, though. As you can tell, I'm from Indie. Sumatra to be exact. I spent nearly three years in a Japanese concentration camp and I have learned something. THE HUMAN BODY CAN LEARN TO ENDURE MOST ANYTHING. So when you have blisters, sore arms, leg cramps or are totally exhausted, DON'T EVEN MENTION IT BECAUSE I DON'T GIVE A SHIT! Worse yet, you complain and I double your exercises. Your mammies are not here to wipe your snot noses anymore, and physiologically you're supposed to be MEN, so here you're GOING TO ACT LIKE MEN! IS THAT CLEAR??? I WANT A UNANIMOUS 'YES SIR!!"

"Yes Sir."

" LOUDER!"

"YES SIR!"

"Sergeant, take over."

The whole gang was hustled outside after that and lined up in columns of three, by height. Being 5'8", Frits ended up at the end of the line. He was nearly the shortest one, just two from the end.

"These are your places. Look at the man in front of you, in back, on each side, know him and fall in in the same spot all the time. Let's number. You call out: 'ONE', you: 'TWO', and so on... " Frits was in row number 27, center.

"ALRIGHT, FALL OUT... FALL IN, DRESS RIGHT, FALL OUT! FALL IN!"

After supper, more haircuts, more trips to the P.X. and Frits sold out his playing cards within the hour. Condoms were not in very much demand, but he'd learned from Joop that it would change remarkably when leave came closer. His pay as a Draftee was going to be 75 cents a day, but here on his first day in the service he already made more than twenty-six guilders. Not too shabby!

He couldn't think of anybody on the floor he really liked, but his bunkie Arnold Fisher, a street bum from Amsterdam, seemed passable. He was fresh as hell, but with a keen sense of humor, so Frits figured they'd get along as long as he kept his things locked up.

The main thing that aggravated him about the boot camp was that they told him how to fold his stuff in his locker. He would have done it neatly, even if they hadn't ordered him to. But by ordering him, he felt like throwing it in a heap.

While most guys were playing cards, he wet the crease of his pants and put them under his mattress for pressing. He shined his shoes and went to bed. It was 10 pm. He whipped out a book and read till "Lights-out".

At 5:15 am, reveille blew. Guys jumped out of bed left and right, but Frits stayed.

Sgt. Van Kluif shook him: "OUT, OUT!"

"I'll get up a little later, O.K.?"

"Oh yeah?" The Sergeant yanked the sheets and blankets off him; "OUT, BEFORE I KICK YOU OUT!"

"You're not allowed to touch me Sarge."

"Oh Boy, listen here, we got ourselves a wise guy. Well, you'll learn to regret that."

He was right. For the next two days, everything the soldiers had to do, Frits had to do twice and he learned fast. It might be easier to play ball with the brass, especially when it was explained he might not go on liberty in 17 more days. "OUCH. Do anything, but don't touch my LIBERTY!"

The marching, drilling, exercising, actually suited him fine. Being a soccer player and a bicycle racer, he was in excellent condition. He was a confirmed non-smoker which was a rarity in those days, and when other guys in the squad were ready to drop, he had enough stamina to start all over again.

That mainly showed up when they started a five kilometer jog in full uniform. As more and more recruits dropped by the wayside, he moved up and up in the column until he was out there out front with Sgt. Bongers and some of the tall guys. When the gates of the KAZERNE came into view, a sprint ensued and he came in second, even beating Sgt. Bongers.

"Next time, I'll sprint earlier and beat everybody!"

"Hey, listen to the little shit over here."

"We'll see."

The sergeant loved it because that type of a competitive attitude inspired others more than speeches or threats.

So even when they went up to 10 kilometers and later on, with full pack and rifle, he was either out front or amongst the firsts, all the time. He learned a lesson. He couldn't sprint with the long-legs. They beat him every time. Also, short distances like 100 meters, 200, or 500, he couldn't win. His chances were at 5,000 and 10,000 meters. There, if he broke early, with at least a lap to go, he could win. If he stayed in the pack he would be beaten by the long-legged sprinters consistently. His legs just wouldn't churn fast enough.

There was talk about Military Olympics in France in August, and the Air Force needed participants. Better yet, real challengers. Now they had his attention!

As the days grew longer, they could practice, racing around the track. And race he did. He wanted to go to France.

Meanwhile, he received somewhat of a bad blow. At home he played soccer for a small, unknown club in a town of just 2,000. In the service, they only wanted guys who played in the big leagues.

It burned him up. When he watched the practices, he knew he could outplay most of the guys, but he was not being considered. He might have been the best player on his home team, but a player on the 15th team of a major club would be chosen because of the name recognition.

So, long distance running became the obsession. Some karate and boxing filled his evenings and left him thoroughly exhausted.

Something else happened that helped convert him into a model soldier. HE VOLUNTEERED FOR SOMETHING! Everyone knows, that in the Military you never volunteer for anything and never raise your hand. If they ask: "Who's a good cook?" and you answer, you get to clean toilets. Known fact. Yet, while demonstrating RIFLE HANDLING, the lieutenant asked: "Who's trained in First Aid?" And without giving it a thought, Frits raised his hand.

"Do you have a diploma?"

"Yes, Sir!"

"O.K., talk to me when we're finished."

Then, and only then, did he realize that he had just volunteered for something. A capital sin according to his brother. Well, this time brother Joop was wrong. After the lesson, the Lieutenant took him to the infirmary where he was asked some basic questions and familiarized with a first aid kit, military style. He received instructions to pick up the kit at 8 am and report to the ammo truck.

While everyone else marched, ran, crawled and jogged all day, interrupted by two hours on the firing range, Frits happily relaxed. He rode the truck to and from the range, sat by the firing line with his kit, and during breaks in the action waiting for new troops to arrive, he fired along with the instructors.

That was one thing he loved: SHOOTING. It made the military more enjoyable. They fired rifles, stenguns, BREN machine guns, and sometimes pistols. He was pretty good at it, and loved the action. That was living!!

The first three weeks had flown by. On Saturday, after the morning inspection, they were headed home for the first time in uniform. After all their BITCHING, most of them were mighty proud to show off their new blue uniforms to the folks back home.

In spite of his "*cool as a cucumber*" attitude, he was so excited he didn't sleep half the night.

At the inspection the sergeant told #2 in row 27: "Recruit, you didn't shave this morning."

"Sir, I have never shaved."

"You have never shaved? Well, I'll tell you what! You report to my room before leaving this post, **SHAVEN!**. You understand?"

"Sir, I don't have a razor."

"Steal one, buy one, borrow one, but you don't go home unless you're shaven."

He was $19^1/_2$.

Military life wasn't so bad after that. All the physical aspects just made his muscles tingle and as life became less restrictive, *JIVING* and *JAMMING* in building B-1-1 became a fun routine.

One of the recruits in B-1-1 played a mean guitar and was the leader of the gang. Frits played drums on the table top and his rendition of "The Chattanooga Shoeshine Boy" became a daily request. Sinatra and Nat King Cole had real competition in that room. Nat's tunes with his trio listed high on the charts of favorites. Everybody joined in when they did *"Oh Give Me Land, Lots Of Land."*

Two radio shows on Saturday afternoons were the main source of their inspiration. One, America's *Your Hit Parade.* The other: *"Swing And Sweet, From Hollywood And Fifty-Second Street."* American music at it's hottest was the most popular amongst the guys.

Every so often a French or English song would slip into their repertoire, but generally *JIVE* and *JAZZ* and *POPULAR HITSONGS* was their real *GIG.* They played and sang their hearts out.

Going out on the town also became one of the greater pleasures. One small club had an "All Dutch" trio, but they were masters of the music of the Forties. Frits would dance his heavy boots off, jitterbugging to *"IN THE MOOD"* and samba like a Latino to the sounds of *XAVIER CUGAT.* All this in his heavy Air Force boots and uniform.

One girl in particular could really swing and when a fast beat would come along, she'd ask him, if he didn't ask her first. She would swing from his left hip to his right, in between his legs and over the shoulders. He was in heaven.

Up to a point. She was about 25 and other than dancing he got nowhere with her. Also, on weekday nights he had to be in front of his bed by 10:30 and on weekends by midnight. That really cut the hell out of his potential love life.

All in all, he wasn't too successful picking up girls in Nijmegen. It was easier to just go home on the weekends and hang out with his old crowd. In Nijmegen, just when he'd think he'd get somewhere with a girl, he'd have to say goodbye and head back to camp. Royal pain!

One sunny day, while sitting on his bench at the rifle range he received quite a shock. A jeep pulled up, he was ordered in, and taken back to the post. The driver couldn't tell him why, and other than stealing food he hadn't got caught at anything. He envisioned time in the brig for stealing, but when he reported to the major's office, he was ordered to change into his "*GOOD UNIFORM*" fast and get back to the office, pick up a ticket and be driven to the station. "You'll have five minutes to spare. Your father was just given his Last Rites! Now move!"

All the way home he thought about his Dad, his Papa. He knew Papa was going to die, that was definite last fall. He suffered from lung cancer and he might not live to see his 56th birthday on July 2nd. Right now, all he could hope for was that his Dad would still be alive when he got home.

The Doctors had told his Mam last November that he might have lived another 10 years if he had not smoked so much or if he hadn't smoked too much garbage. Three packs a day of unfiltered garbage, the worst he could find. Black French, homegrown, *ERZATZ* during the "*war*", anything that burned! What a waste!

Frits and his pap had a peculiar relationship. Some friendship, some love, some respect and some disrespect. His Pap had an explosive temper and his ulcers made that even worse.

It came to a halt, when Frits told his mother that his brother Joop at age 17, was too old to be slapped around for snitching a slice of baloney. When Papa heard the commotion he came running into the kitchen from behind the bar and when he saw Mama crying and Frits shouting, he grabbed a poker from the stove and clobbered Frits smack over the head.

To this day, Frits wasn't sure whether his arm kinda broke the blow or if he just had a very hard head, but he jumped up, wrestled the poker away from his Dad and told him he'd kill him if he ever laid a hand on him again.

It all ended with a lot of tears and promises, and later Papa tearfully told him how somehow his children had grown up unnoticed and how he and Mama still looked at them as "our little kids!" All that would change and it did.

Involuntarily, Frits felt for his scar between his hair. Ever since that fight, he and Pap had become good friends and it was a shame to lose him that soon.

He didn't lose him. Not yet!

After seven days and several phone calls and telegrams

from the Army and the Air Force, Pap's doctor gave them a "slip" and Joop and Frits returned to their posts. Papa was still hanging on.

Shortly before boot camp was over, several guys from Building "B", (the singing room), O.K.'d for flying status, were hustled out to go to Flight-training. One of them, "Hompie" van der Heide, left in handcuffs because he was serving time for something ... Frits didn't know for what. He also didn't go. When he asked, he couldn't get an answer.

A week later boot camp officially ended and everyone left for their assigned post ... everyone except Frits. He was moved in with the drivers and mechanics of the "*STAFF*" and was told: "Don't worry, just help out here and there."

In a way it was great. He didn't have to report in by midnight and that greatly improved his dance habits.

Besides that, the *"VIERDAAGSE"* started up. Each year, people from all over the world come and march in and around Nijmegen for four days. It is the highlight of the year. The whole town celebrates. It's like a Carnival, or a Mardi Gras.

Each morning the marchers line up and start their trek through the hills and lowlands of Holland. Air Borne troops from America, Scottish Pipers, Tommies, Youth organizations, Tyrolers, Freelancers, literally thousands and thousands.

As they return at night, throngs of people cheer them on as they march the same route that was marched by the ROMANS, NAPOLEON, the GERMANS, the BRITISH and AMERICAN TROOPS. What a sight!

Frits was working as a volunteer and sat by a rest stop with a Red Cross truck, treating blisters and sunburn.

A sharp looking group of Israeli soldiers in shorts and high-quarter boots, their socks rolled down over their boots, stopped for a break and Frits treated one of the prettiest blisters on the daintiest foot, attached to a beautiful leg. The group consisted of female soldiers.

Her name was *"Leva"* and her German was flawless and so was Frits'. By the time he gingerly put her sock back on, he knew where he could find her at night.

She was nearly nineteen, was drafted and loved the military. She was also looking forward to seeing him at 5. What a beauty, what a face! Dark curls framing a suntanned pretty face. But the eyes, *THOSE EYES*! Liquid eyes, nearly black.

At first he thought that she was hurting, that tears welled in her eyes, but that was their natural look. *He was in love!*

He mended many more feet, nice feet, indirectly attached to nice faces, but LEVA's was the only one he could think of. He nearly hit a snag. When they were informed that stragglers would keep them for an extra hour or more, he nearly panicked. A very understanding sergeant-major let him leave at 4 anyway, 'cause he understood how important it was that Holland and Israel established good relationships.

He easily bummed a ride into Nijmegen and found Leva as planned. Unlike Frits, she was allowed to wear civilian clothes and he had to wait a while for her to change. But BOY, was it worth it! She had changed into a fairly short skirt that made her look as if she was only 17, but OH so feminine."

Her foot still bothered her a little, so she would rather not go dancing, but would rather go see historic Nijmegen. She was just great. She drank beer! A leisurely beer in a sidewalk café , a tram ride to the "VALKENHOF", and another leisurely beer, leaning on a wall that was built by CHARLEMAGNE in the late 8th century.

She smiled at Frits' enthusiastic explanations as if he were a trained tour guide showing the sites to a customer. She probably wasn't listening at all, because in the middle of one of his sentences she interrupted and with her face close to his, she asked:

"Where did you learn to speak such beautiful German?"

He just leaned over and tenderly kissed those full young lips.

When five minutes later, their bodies close together, one hand around her neck, the other around a beer, they broke the kiss, he murmured softly into her cheek:

"We were occupied by Germans for five years. I grew up with two languages, how about you?"

"I grew up in Austria. My native tongue is Hungarian, but an Austrian family brought us up. After the war an uncle of mine from Israel came and got us."

She pressed his lips back to hers.

"Let's get away from all these people." Teenage love overcame the fascination of a twelve hundred-year old restaurant.

Boot camp had concluded four weeks ago, the Olympics

in France had passed and because Frits had no "SQUADRON AFFILIATION," he was not invited.

"DAMN!"

Finally ... *"Report to WEESP!"*

"WEESP? What's in WEESP?"

"A warehouse. Here are your orders."

"A warehouse in Weesp? My uncle lives in Weesp, my father's youngest brother. It's close to Amsterdam and I can bring my bike down and make it home in an hour."

What a crazy deal. He worked in a Military Supply House. He learned to spray paint. Started at 8, worked till 5. It was like a civilian job with less pay.

Not a bad deal, crazy maybe, but not bad. He figured he could survive this for one year and he'd be out. Discharged!

Some nights he'd go into the big city, Amsterdam. Some nights he'd have dinner with his uncle and aunt and their son Eric.

Friday nights, he pedaled his way home. Easy living, he could get used to that!

CHAPTER II
MECHANIC?
"WHAT NOW?????"

He'd been called to the office.

"Here are your orders. You're going to "DEELEN." "MOTOR MECHANIC SCHOOL." "M.M.2"

"Motor mechanic? I don't wanna be a Motor Mechanic. Dirty, filthy engines. Who the hell dreamed this up? Can't you get that cancelled?"

"Have a nice trip!"

There he was ... CAMP DEELEN. Huge Air Base, hundreds of bunker-like buildings, built by the Germans. They slept and ate in the heavy-walled buildings and they attended classes in the bunker-like buildings.

This was the dumbest thing the Air Force had come up with so far. After half a day in class, he asked to appear before the SCHOOL COMMANDANT. The next day he appeared before Captain/Pilot/Navigator Woudenberg.

"Yes Sir! Thank you Sir! I'm sitting in a class, listening to stuff I learned seven years ago. I know that stuff better than the Corporal who's teaching it and I can probably teach it better than he does. Besides that, I don't want to be a mechanic!

"Just a moment, soldier. Yes, I see you have an extended education, M.U.L.O., A plus B, so I believe that you know more than the Corporal when it comes to math. But tell me, what would you like to do?"

"I want to be a pilot, like you!"

The Captain chuckles. "So why don't you apply for it?"

"I did, Sir."

"Well, were you rejected?"

"NO Sir, I passed everything and now I'm here."

"Hold it a moment, soldier. You passed the physical at Soesterberg?"

"Yes, Sir."

"And the two days at Ypenberg?"

"Yes Sir!"

"Than what are you doing here?"

"Sir, that's my question. What am I doing here?"

"Just a minute. SERGEANT! Get this guy something to do for a few days and have him come back!"

"I'll see you in a few days, Forrer."

"Thank you Sir."

After a few days, he received a message from the Captain. "Assignment confirmed. Will start Flight training upon completion of M.M. 2 training."

"They want me to finish this stinking course first and then I'll start Flight Training? By then I'll have more than a year in the service. The hell with them. Once I have more than a year in, I'll survive the other few months and I'll be out. The HELL with them!. I'm gonna fail this course and get out. Screw them. I don't have to live with this screwy organization. I wonder who runs this mess?"

Until his class finished the preliminary math classes he was assigned to warehouse duty where he drove everybody crazy with his "PERFECT" imitations of Johnny Ray. Not only could he do great renditions of *"CRY"* and *"A Little White Cloud That Cried"*, but he could do them louder than Johnny Ray ever could. His comment: "These clowns don't appreciate good music!"

Finally his class, totaling ten, started engine mechanics classes. Much to his surprise he found it fascinating. First: Piston engines. In-line and radial. They tore apart and assembled engines for the T-6 and an "AUSTER". Next: Rolls-Royce engines for the GLOSTER-METEOR. Real JET engines. Intriguing!

While most kids studied hard at night, he'd ride his bike to Arnhem to visit his Tante Paula and her kids, or he'd go dancing somewhere. Some weekends he'd hang around Arnhem and sometimes he'd hitchhike home.

One weekend he caught a ride with a lady who drove a Hillman station wagon full of office supplies. The lady was very pleasant, about 30 and married. Her husband was a career officer in the Army, stationed in Indonesia. She told him she worked because she had no children and enjoyed working and traveling.

She smiled benignly when he told her he was going into pilot training after he finished engine mechanic school. He could feel she didn't believe him.

"Tough," he thought. "You can believe whatever you want." He didn't say anything, though. After all, she was getting him halfway home.

Along the way she had to make a call and when they pulled into a restaurant parking lot she suggested: "Have a beer, I won't be long."

Shielded from the wind by a glass partition he enjoyed a cold beer on the open veranda. When she came back she was in no hurry and had a beer as well. She asked about home, what time he was expected and when was he to report back to base.

He explained that he'd have to be in at midnight on Sunday and it being Saturday at about 1 PM, that didn't give him much time. She wholeheartedly agreed. That did not leave much time. What time did his folks expect him?

"My folks don't even know for sure that I'm coming. Some weekends I do, others I don't go home. When there are birthday parties or "KERMIS" (Mardi Gras), I go home for sure. Otherwise, I hang around Arnhem or go to Nijmegen. All depends."

"Have you ever been to Loosdrecht, Frits?"

"The lakes?... Sailing?... No, I've never been there."

"Well, it's too cold for sailing but it's only twenty minutes from here. Want to take a look?"

"Sure!"

"Do you want to call your folks, tell'em you'll be late?"

"No, that's alright, they'll assume I stayed on base or in Arnhem. No sweat."

"O.K., let's go. No, no, I'll pay for the beers. I know how little you guys get paid."

Later on he found out that she had called to make sure a little lakefront cottage was available and that's where she took him. She bought beer and an electric heater. She told him about her lonely life without her husband and the confinement of living amongst friends and relatives, who expected her to sit home and knit while waiting for his return.

The cabin was cozy and small but the heater didn't give off enough warmth to warm the cabin, so MARION, as she was called, had him under the blankets before six. First he was too wild, but then she started to teach him. Tenderly, slowly. Great teacher. She laughed when he told her he had never been in bed with a woman before. On the beach, YES. In the park, leaving stains on his brother's trench-coat, YES. Standing up, in the alley behind the house, YES. Underwater in the canal, YES. But at his leisure in a bed? NO.

Well, she would teach him. And she did.

By eight o'clock they washed up some (there was no shower), and went out to eat. A cute little place that couldn't hold in the crowds during the summer was fairly empty at this time of year. It left them the dance floor all to themselves. Marion kept complimenting him on how well he danced and in turn he told her how he learned to waltz with his Mam when he was only six.

She learned to rumba and the slow waltz, which she learned was called the English Waltz. She just loved it. They had a ball.

On the drive back to the cabin, she told him: "I'm going to buy you some civilian clothes. Those boots of yours have got to go!"

"I have clothes, but we're not allowed to wear 'em. We're at war, you know."

"I know, my husband's in it, remember? But when you're here no one will know you're in the service, so don't worry."

The bed got quite a workout in the little cabin that weekend. They didn't get up till ten and then mostly to eat something before crawling back in bed. Frits had never known such a wide variety of things to do with a human body, or how to do them. They went from the most tender to the most erotic kinds of lovemaking.

By noon, just lying under the covers, touching and sipping beer, she made him vow two things: "Don't tell anybody about this, NOBODY! O.K.? This is our own scandalous little secret, so NOBODY, ABSOLUTELY NOBODY. O.K?" He just nodded!

"Let's face it, I need this and I love doing it with you, but when my husband comes back, I want to continue my life with him, you understand? The next thing: please don't fall in love or get attached to me. You're young. You eventually need to find a girl your own age and settle down with her to raise a family. Meanwhile, this is just fun for both of us, O.K.?"

"She's using some big words." Frits was thinking, "LOVE?" I don't think this is LOVE!. This is good, raw unadulterated SEX. And that is fine by me!"

But he said: "A deal. We get together when we can and make the most of it. Meanwhile, let me get back at that nipple! You're beautiful, you know?"

She drove him back to the base by ten. He walked like a cowboy, but he slept like a baby.

A week later he was to meet her again, but instead he was home. Another telegram: "LAST RITES".

They again stayed home, day after day, sitting with Papa around the clock, watching him cough out his lungs.

On November 2nd, 4:00 AM, he mercifully died. His weight was down to 85 pounds. Not until the funeral, as the casket was lowered and Joop and Frits snapped to a final salute, did it finally hit him: "PAPA IS NO MORE!" He fought back the tears, but it wasn't easy.

He tried to get with Marion for the week-end, but her mother-in-law was over, so she couldn't make it until the next Friday, when he would be off for a long week-end. She would pick him up and they would have three nights together: Friday, Saturday and Sunday. He really needed her "NOW", but it would have to wait.

When he did get back to the base he was immediately hit with some bad news. His squadron had failed a major barracks inspection and because of that, his squadron had drawn guard duty for that week-end and he was restricted to the base.

"WHAT A BUMMER! WHAT A GODDAMN BUMMER! And I wasn't even here. Now how can I get blamed for this screw-up? I won't stand for it and I'm going to ... "

Well, he really couldn't do anything about it and he was about to call Marion when the Guard Roster came out. He wasn't on it. He was on Special Leave when the roster was made up and nobody knew for sure when he would be back, so they hadn't scheduled him. He decided: *"I'm simply going to leave."*

Friday afternoon, after final inspection, he marched through the gates to the waiting busses that brought the boys to the railroad station where he would wait for Marion. His heart was in his throat all the time. When he walked through the gate he thought someone would snatch him right there. On the bus, he kept his eyes peeled for N.C.O.'s that might know him and when he saw the M.P.'s at the station he thought: *"This is it, they've been alerted!"*

NOPE!, NOTHING!. Nobody cared about him and he calmly walked across the square to the pub where she would meet him. He toyed with the billiard balls and was just on his second beer when she walked in.

She seemed prettier and younger and he had a hard time acting casual in public. She drank a quick beer and they were off. The moment she was behind the wheel and had the car started, she leaned over to kiss him and he snuggled as close to her as the gearshift poke would allow him. Wrestling with the stick-shift, he slid his hand into her blouse, hungrily kissing her. She reached over and when she felt him rising within his pants, she squeezed him hard and said in a hoarse voice: "Let's get out of here!"

She hauled out and within five minutes pulled into a dirt road, out of town. She turned off the light, but left the engine running so it would keep blowing heat. She crawled to the passenger side, straddled him and rode him to an explosive crescendo. After that, it was as if the excess pressure had been relieved from a steam engine and they settled down to a leisurely ride to the cottage for a leisurely erotic week-end. THE BEST WEEK-END OF HIS LIFE. It also proved to be one of the worst.

At 7:00 AM on Monday morning he strolled into camp. At 7:30 he lined up in formation and by 8 o'clock he was at his desk, learning about Jet-Engines.

Nobody said a word. Nobody seemed to have missed him. By 10:30, a soldier from Administration came in and, apologetically, verified his presence against the attendance list that the instructor had sent off, just after 8.

The young soldier saluted: "Thank you, Sir" and left. Frits sighed a sigh of relief.

NOT FOR LONG!

At 11 AM, two big M.P.'s reported to the classroom with orders to arrest Private Forrer.

Out he went! In Captain Woudenberg's office he caught hell!

"What possessed you? Absent Without Leave for 62 hours during war time? Do you know that the minimum is THREE MONTHS IN NIEUWERSLUIS? (Leavenworth). Do you know what that means?"

"No Sir."

"That means three months in maximum security. Then after your service time is over, they'll add three months. Forget about promotions, forget about a Flying Career. That's over! What the hell were you thinking? Didn't you know you were restricted to the base?"

"No SIR!"

"No Sir? Don't you read?"

"I wasn't on the list, Sir."

"What list?"

"Guard duty roster, Sir."

"So you mean, you weren't scheduled for guard duty, so you just took off?"

"It was a long weekend, Sir."

"But you were restricted to the base, your whole squadron, for failing inspection the week before."

"I wasn't here, Sir."

"You weren't here? When and why not?"

"I was at my Father's funeral, Sir."

"YOU WHAT?"

"I was at my Father's funeral, Sir."

"When? Skip that! I'm sorry to hear that son." His voice mellowed and he slumped in his chair.

"When did you bury your father?"

"November 5, Sir."

"Guard, take this man to the Guardhouse. Give him some coffee and wait for my call."

"Yes, Sir!"

After an hour, there was a call for him at the Guardhouse.

"Frits, this is Captain Woudenberg. Listen. I've talked to Colonel Wiersema, the Base Commander, and we'll have a hearing in his office at 9 AM tomorrow morning. You'll stay in custody till that time. I'll pick you up at 8 AM. Be ready. Spit and polish. I'm going to defend you so don't worry too much.

Hang tough!"

"Yes, Sir. Thank you, Sir."

"Give me to the Duty Officer please."

"Yes, Sir."

The two big M.P.'s took him to his quarters to collect his GOOD uniform and toiletries and escorted him to the brig. Same heavy brick, German built building with six cells. One occupied. Frits made two.

The voice in the other cell was lonely. He talked or asked incessantly. Frits wanted to be left alone so he could dream.

The cell was bare except for a wooden bunk. It had a little window high up on the wall and a solid wooden door with a little bitty window in it that stood open. That was it. It would have been more peaceful if that other clown had stopped talking, but he learned to close his ears to it so his imagination would have free reign.

He tried to evoke feelings of guilt or regret, but it had been such a marvelous weekend that it seems worth it, whatever the penalty might be. One thing worried him though. How would he get to communicate with Marion? She might not be able to find out what happened. "Oh well, keep dreaming", he told himself.

At 8 am, the Captain picked him up and took him for coffee. He instructed him to say nothing but "YES SIR" and "NO SIR" and look remorseful.

At the Colonel's office, with a Major from the Court Martial Board present, Captain Woudenberg submitted a plea that astounded Frits, but that worked.

"NOT BEING ON THE SQUAD ROSTER, NOT HAVING BEEN PRESENT DURING THE WEEK-END THAT THE SQUADRON DEFAULTED AND OVERLY DISTRAUGHT OVER HIS FATHER'S RECENT DEMISE, HE HAD NOT INTENTIONALLY GONE A.W.O.L., BUT HAD THOUGHT IT RIGHT TO GO HOME AND HELP HIS MOTHER WITH HER BUSINESS DURING THIS DIFFICULT TIME. THERE WERE FOUR YOUNGER BROTHERS AND SISTERS HOME, THE YOUNGEST BEING ONLY FIVE AND ALL CLOSE TO THEIR FATHER AND NEEDING FRITS' SUPPORT. HE HAD REPORTED BACK, RIGHT ON SCHEDULE AS HE UNDERSTOOD IT. HE HAD BEEN SELECTED FOR FLIGHT TRAINING. HE WAS A FINE YOUNG

MAN AND CHARGES SHOULD BE DROPPED."

The major said a few things about "*HIGH TREASON DURING WAR TIME*", and Frits was dismissed from the room, to be recalled a few minutes later.

The Colonel, a Pilot also, announced that they were having *EXTREME MERCY* on him and sentenced him to the brig for 14 days, starting today.

"Yes, he could call his mother."

As the Captain drove him back to the brig, he said: "You are extremely lucky." Apparently they liked the word EXTREME.

Back in jail he asked the guard to turn the radio to the A.F.N. (American Forces Network) and leave the hall door open so he could hear it. He was allowed paper and pencil and as fast as he could write, he wrote down the lyrics to the NEW hits by that NEW star, Tony Bennett: "*Cold, Cold Heart*" and "*Because Of You*". It was rare that a new singer hit the charts with two HITS, but this guy did it.

By nighttime, he knew the songs by heart and he'd sing them for his unseen neighbor. The neighbor didn't speak any language but Dutch, but he appreciated any and all noise that Frits produced.

The next morning the Chaplain came to visit and volunteered to call his Mam for him, but that was not the woman he wanted to call and he needed to time his calling privilege precisely. He didn't need the Padre for that.

When he told the Chaplain though: "Today, my third day in the brig, is my birthday, my 20th", the Padre got sentimental and insisted on praying with him. Well, at least he left him some good books to read and promised to be back. That was nice.

His lonely neighbor was a little slow in learning a game called "BATTLESHIP", in which, within a hundred squares, drawn on paper, 10 X 10, you'd draw a flattop, cruisers, battleships and subs. Then you'd fire at one another by calling of the numbers of your grid. Sometimes they'd play for hours and sometimes, Frits would read for hours. Until "UNSEEN LONESOME" would call out: "*SING SOMETHING, IT'S SO QUIET!*"

He'd get to enjoy a concerto, a cappella, only accompanied by knuckle drums on the bunk. It might be a half hour of DORIS DAY or an hour of WALTZES. Sometimes it was a program of the ANDREWS SISTERS with DANNY KAYE or BING CROSBY,

Motor Mechanic–2nd Class

and sometimes nothing but "LATIN" from MARIA ZAMORA or XAVIER CUGAT. SINATRA could fill an hour all by himself and German marching songs accounted for at least two. Dutch songs could go on for hours on end and French songs, new and old, would easily fill an hour.

The days went by fast. Sometimes a Sergeant of the guard would not appreciate "*A Little White Cloud, That Laid Right Down And Cried*" or "*There Was A Time When I Believed That You Belonged To Me*" and made him shut up. That's when he delved back into the sexy books that the Padre had brought him. He never knew that NAPOLEON screwed around as much as he did and that so many whores followed him all the way to Moscow.

It was an education, gone by in no time.

Marion picked him up when he got out one cold Monday night, but he had to be in by 10:30 and the following weekend she was not available. Frits went to see his Mam. It had been a month since he'd seen her. She was holding up alright!

Classes were a mixture. Engines were interesting but working on them was boring, dirty and cold. He made up his mind. He'd NEVER become a mechanic. His teacher was a charming, fatherly type of man with whom he got along well, and for whom he translated English technical texts when the teacher got stuck. It was a pleasant arrangement.

As the exams came closer and graduation was not too far off, his determination grew. HE WAS GOING TO FAIL. He would have some six more months to serve and he would be out. Good-bye Air Force!

Several of his classmates felt the same way or at least said so. He was not too enthralled with the whole military establishment. Too much brown-nosing to gain or keep their jobs, too much unnecessary chicken-shit, all in all, too many things against which he revolted.

He didn't know what he wanted to do with his life, but being a part of a big company or the military was not it. So when in March, the final tests were taken, he failed. Four more in his class failed and three others had to retake part of the tests. ONLY TWO PASSED! The instructor was thunderstruck!

Frits transferred to The Hague. "THE OLD ALEXANDER KAZERNE". An ancient building that housed the low-ranking staff of Air Force Headquarters in Scheveningen. It was a residence where guys ate and slept. Period.

At 8 am, they reported to work somewhere and at 5:30 they ate if they wanted to, and at midnight they had to be back in. Week-ends, they could do as they pleased most of the time. If they wanted to eat on the week-ends, all they had to do is register for it. It seemed plenty simple.

Frits' main job was that of a messenger. He pedaled around on an old Air Force bike from one office to another. He'd pick up papers here, deliver them there and flirt whenever he could. Life was not very purposeful, but easy. On week-ends he rode his bike home many a time and sometimes he'd stop in Haarlem to see his Aunt, Tante Engelien, his father's only sister.

The Hague and Scheveningen had some exciting Jazz Bars

and Nightclubs and he struck up a friendship with another Jazz-Fan in the kazerne. JEFF, he pronounced it Sjef, was a trumpet player of sorts and he showed him around to some of the Jazz joints that they could afford, and to some that they couldn't.

They saw "ELLA" in the CONCERTGEBOUW in Amsterdam, Cab Calloway in some dive in the same city and Rita Reys with Nellie Lutcher in Scheveningen. Later on they took in the Divine One, Sarah Vaughn with Duke Ellington and on and on and on.

Many week-ends he worked at home as a bartender or waiter or for some of the other bars in the area. Just to earn enough money to pursue his Jazz interests. Jeff was not much of a dancer, but Frits would dance his shoes off and sing his heart out every chance he could.

In their dorm, 12 cots lining the walls, they'd jive and sing. Some of the garbage cans had good sound and did beautifully as bongo drums. Nat Cole's *"That's My Girl, Take A Look At Her, She Belongs To Me"* provided easy swing and *"I'm Forever Blowing Bubbles"* would go so fast that Jeff's trumpet could hardly keep up: *"Boobleabob, Boobleabop*!" Julie Christy's '*How High The Moon"* was their COOL tune and so they spent endless evenings jazzing away.

When something was happening and they had money, they were out there, being a part of it. When they were broke or it was a dead night out there, they would just JAM the evenings away and have a ball.

One night Jeff asked him over a beer: "Why is your name so familiar to me?"

"Because I live in the same room?"

"No, no, that's not what I mean. I've seen your name before somewhere before you ever came here."

"I don't know", Frits was kidding him, "Maybe in the movies?"

"No, seriously. I work in the office of Captain Kruis and we handle Flight Training. Has your name ever come across my desk?"

"Possibly, because I was tested for Flight Training about a year and half ago."

"Oh? And you were rejected, right?"

"NO WAY! I PASSED EVERYTHING WITH FLYING COLORS!"

"Then why aren't you flying?"

"They made me finish this M.M. 2 course first and now, so much time has gone by, I'm no longer interested. I have four more months to go and I'll be a civilian again and that's really all I want. I'm sick of the military."

"Really? I'm gonna look you up though. I file the records, see. They all come through my fingers, so I have immediate access to them."

"Check it out. You'll see I'm telling the truth."

"I will."

Meanwhile, the Service started to aggravate him again. For the May 5th celebration, the anniversary of the German surrender in '45, the Air Force was going to participate in the parade down the boulevard along the beach in Scheveningen.

Result: Drills at night, trying to get the pencil pushers and delivery boys back into some sort of military order again. They're supposed to resemble a marching unit. He hated it. He had planned to watch the parade and the Queen from the sidelines, in the company of a pretty blond Jazz enthusiast who enjoyed his dance instructions.

She wanted to learn the Jitterbug "MOKUM" style (Amsterdam's way), as well as "SWING" Artie Shaw or Benny Goodman style. They had danced a few times in some of the pubs, but this was to be their *first* date. Tentatively, on May 5, they were to go to the parade together and RIA, a nursing student, with little free time for herself, was off that day. *DAMN, DAMN, DAMN.*

She was an intern at a Catholic Hospital and *BROTHER*, these Nuns were strict. This was to have been their first full day together and then the stupid Air Force pulled this kind of crap.

"DRESS RIGHT ... DRESS! FORWARD... MARCH!.... Hut ... two ... three ... four!"

"Hate it, hate it, hate it! Can't wait till I get out!"

Thank God for Guardian Angels.

On the eve of the parade, he came up with a badly infected tooth and cheek, puffed up as if he were hiding a prune in it. In the clinic he received some pain pills, but they couldn't schedule a dentist till 8 AM on Saturday morning. *"HALLELUIAH!"*

"Sorry Sergeant, I have an emergency surgery at 8 AM.

Sorry." "*HALLELUIAH.*"

He might have a sore tooth but he'd be watching the parade rather than: *"FALL-OUT at 7 AM., STEP-OFF 7:30, arrive parade area 8:30, ASSEMBLE 9 AM, STEP-OFF for parade 10 AM, finish parade 11, back at the old Alexander 12:30 PM. POOR BASTARDS!"*

"I'll wave at you, schmos. When you see me with a blond, just blink, don't smile. I would have to report you! You poor bastards, don't worry though. I'll have an extra beer or two so you guys won't feel so bad. Poor bastards."

At 8:05 the dentist fished a little, 5 mm piece of pork-rind from behind his lower gums, rinsed and cleansed the area and said: "This was the culprit, this got you infected." He gave him some antibiotics and painkillers and told him to come back in a week if he still had a problem.

At 9, he picked up Ria at the hospital and they headed for the beach holding hands as they were riding their bicycles. The swelling in his cheek receded rapidly and the pain was negligible. So out on one of the many terraces lining the boulevard he took off his jacket, enjoyed the cold beer, the warm sun and a golden girl. What a way to live!

May 5th was also his mother's birthday and he needed to call her, but while he was under the influence of a pair of limpid blue eyes he figured he'd get around to it later sometime.

When Ria commented on the brilliant blue skies and the warm spring sunshine, he told her that on his mother's birthday the sun always shone, according to his Mam. She was born 5-5-'05 and she claimed that only once it had rained a little on her birthday. Ria then insisted he call immediately. He was glad he did, but his Mam wanted to know if he were coming home and he stalled her with "*THE PARADE AND EVERYTHING!*" and wondered if he should bring Ria home with him. Tell Mam: "*HAPPY BIRTHDAY*, and I'll see you later."

The parade was impressive. Tanks, armored personnel carriers, artillery and marching units. The Air Force Units on their rubber soles, glided by as if they were wearing sneakers. He waved at them as they "shuffled" by, compared to the Army units that came "clanging" by with their metal-tipped boots.

He got a serious "pang" when the fly-by came. T-6 HARVARD trainers, Navy SEA-FURY's, but the worst was when 16 GLOSTER-METEORS came roaring over. He squeezed Ria's hand and said: *"BY GOD, I could be doing that!"*

"Why don't you?"

"I got turned off, I guess."

"Can't you still do it?"

"Oh, I don't know. Sometimes I would just love to fly one of those and most of the time, all I want is to get out of this lousy uniform."

Jeff brought home Frits' records one night. "I don't know what happened. You were supposed to start training last year in July in Hilversum. But I wasn't there then, so I don't know what happened. Maybe somebody just mislaid your file. Then in September there's a notation: "Start training after M.M.2." and I have no notation that you finished it."

"I did. I failed."

"But you finished, right?"

"Right."

"Well, where do you want to go? HARVARDS in Gilze-Rijen?, then GLOSTER-METEORS in England, or would you prefer TIGERMOTHS in Ypenburg, Hilversum or Woensdrecht,and then America for F-84's? Which would you prefer, Sir?

"Sure, as simple as that, right?"

"Yes, as simple as that. That's what I do, I do the scheduling."

"You're kidding!"

"I'm serious!"

"Oh hell, I don't know. All I can think about is getting out."

"Well, kind Sir, let me know."

"O.K., now get your trumpet."

"OH How I Hate To See That Ev'ning Sun Go Down ... Pahboobababoo ... O, How I Hate To See That Ev'ning Sun Go Down ... Blow It Brother!" Tonight We're Doing The Blues ... "Ol Man River, That Ol Man River... " 'Yea man, blow that there horn ... teach Gabriel how to blow that horn ... ! Yeah man" ... singing the blues all night ...

Then something happened ... He was called to see his squadron commander, a skinny 1st Lieutenant with buck teeth.

"At ease, private. I have orders here from headquarters. It has been found by the EXAMING BOARD, that inasmuch 7 out of 10 students in class M.M.2-81 failed their final exams;

whereas, out of 141 students graduating at the same time, only 11 failed that class M.M.2-81, willfully and deceitfully failed their finals. Therefore, the 7 aforementioned will serve their 6 months extension as if they had graduated. Signed ... etc."

"Attention, Private! Dismissed!"

"But Sir ... "

"Dismissed!"

"Damn, Damn, Damn. Those Goddamn lousy bastards ... Damn, Damn ... "

"Jeff? Frits here. How does America look?"

"O.K. ... O.K. !!!"

That evening Jeff swore he had it all fixed. Orders cut and signed. Ready to go.

Next morning train ticket to Bergen op Zoom, bus ticket to Woensdrecht. Orders: "*REPORT FOR FLIGHT TRAINING!*"

"HOT DAMN! FOR REAL? ... HOT DAMN!"

CHAPTER III
THE TIGERMOTH

"Have you ever been in a plane before?"

"Not when it was off the ground."

"Not even commercially?"

"Nope, never been off the ground."

"Are you ready anyway?"

The 1st Lieutenant was a roly-poly friendly man with laughing eyes. He should be fun.

"As ready as you are, Sir!"

"Alright, strap your ass in the backseat. You've been taught the use of the parachute, right? Good, you might need it."

"Really?"

"Yeah, because if you don't tighten those shoulder straps any better, you're gonna fall right out of the damn plane. That's alright though, just wait till you have enough altitude so you won't break your neck when you hit. Now just sit back and enjoy!"

He pulled his fat frame up the side of the little yellow biplane and lowered himself into the front seat.

"AF, OPEN, DICHT!" (magnetos off, gas on, throttle closed) he hollered at the mechanic out front, who swung the propeller a few times. Next a "THUMBS UP" sign, "CONTACT," one more swing and all 100 horsepower kicked into action.

Lt. HOXEL gave it some throttle and started zigzagging

down the grass field. Before lining up between the "DOGHOUSES" that marked the runway, he faced the engine into the wind and ran-up the engine some. Seemingly satisfied, he rolled in between the DOGHOUSES, looked over his shoulder and hollered: "YOU O.K.?"

"O.K.!"

"Put your hands on the stick and feel with me!"

He talked into a tube that was connected to Frits' World War I leather flying helmet and without waiting for an answer, he shoved the throttle forward and they went bouncing down the grass strip. It didn't take very long before the tail came up and he was level with the instructor in front. Now he could see forward as well. A few more bumps and they were OFF. *THEY WERE ACTUALLY FLYING*!

For the first time in his life, he was actually flying and climbing higher. The sensation was not all that unusual. It felt like riding in a car, but suspended.

What was elating was the thought that he was actually flying. Twenty years old and flying! Unbelievable! No one in his entire family had ever been up in an airplane before and here he was, climbing toward the clouds and watching the world from up above. All those neat patches of green below, dark, medium and light green, pastures, woods, grain fields and potato patches, all nice and orderly and ... "*OH SHIT*!" He hung upside down suddenly. It happened so unexpectedly that he grabbed the cables along the insides of the plane.

It startled the living hell out of him momentarily, mostly because it was so sudden. The voice in his helmet said: "I told you to tighten those belts and keep your hand on the stick."

"How did he know that?"

The engine stopped running and the plane rolled back to level flight. They glided noiselessly for a few seconds and the engine kicked back in. The whole thing had taken only a few seconds, but Frits found himself tightening his belts more securely. For a moment he had thought that he was going to fall out of the plane.

Right then and there they were diving down to earth, the engine roaring at full speed, then up ... and up ... and over ... upside down again ... but without the sensation of falling out. The plane leveled out and the helmet told him: "That was an IMMELMAN."

"Whatever you say!"

"You O.K.?"

"I'm O.K."

The chubby guy in the front seat pulled the plane nearly straight up. For a moment it seemed like it just hung there, suspended ... then ... SNAP, the nose slapped on down and the plane turned 'round and 'round, going straight down to ground.

It gracefully stopped it's turns and with a pleasant roar pulled out of its dive and then, the engine blasting again at full power, they were climbing back up.

"That was a spin. You like it?"

"That was fun!"

Once again, he dove down the little yellow aircraft at full throttle and once again he pulled it up and up and up and this time they went over the top, the nose going through the horizon, upside down and then straight for the green pastures.

As the little bi-plane pulled out of its dive, he was pushed hard into his seat, the power was pulled back and when they got back to a level attitude, the engine started racing again. Once level, the voice in his ears asked: "Do you know where the field is?"

"Oh, hell ... " At first when they flew out he knew they were flying north-west toward Zeeland, but ever since they started clowning around he hadn't given it any thought. In the distance he could see the silvery slivers that had to be the OOSTER and WESTER-SCHELDE and behind to his left was a large town that had to be BERGEN OP ZOOM.

"It must be behind us Sir."

"O.K."

He laid the plane in a left bank and when he leveled off he asked: "Can you see it now?"

"Out in front and to the left a little more."

"Good show! Put your feet on the pedals and feel with me."

The instructor talked incessantly, explaining every pressure with his feet, every movement of the stick, every forward or backward motion of the throttle, and Bump ... Bump ... Bump ... they were on the ground again.

"How'd you like it?" Chubby was undoing the parachute straps.

"That was fun! That was better than a roller-coaster."

"Good, give me your logbook"

"What's a logbook?"

"Go into the ground school and get a logbook. You're in the flying business, soldier."

HIS FIRST FLIGHT LASTED 19 MINUTES!

His excitement hit a temporary snag. He signed a contract, becoming a "RESERVE SOLDIER," obligating him to three years if he completed pilot training. If he "WASHED-OUT," he'd go back to "REGULAR" status and duty and the contract would be null and void.

One advantage was – his income doubled. From ONE Guilder per day to TWO per day. That sounded great. BUT, there was a big BUT. Since he was not a DRAFTEE anymore, but a VOLUNTEER, his mother's Government Support for financial hardship was cut entirely.

"DAMN!" He hadn't told his mother yet that he had transferred to a Flight School, and now this. There was one thing he could do ... he could sign a voluntary transfer. That way, once a month, his Mam would get his extra guilders and he would still have his ONE Guilder a day. WHAT A BUMMER!

But it was better than nothing, and he signed, remaining as poor as he had been before, but happy just the same.

The Air Base was nothing but a great expanse of grass with a bunch of buildings at one end. The buildings and the base were built by the Germans and from here, close to the North Sea, the Germans would scramble their planes to meet the incoming ALLIES.

Of course the base has it's ENLISTED quarters, its N.C.O. and OFFICERS quarters, separate Mess and clubs for all those levels of rank, a movie house and classrooms.

It was a training camp all over again with inspections, early rises and bed-call at 10:30. The atmosphere though, was much more relaxed than it had been in boot camp and with much more camaraderie than in Motor Mechanic school.

The other student pilots were educationally and socially definitely his equal and in many cases his superior. There were many more "music lovers: and there was more of a "devil-may-care" attitude amongst them. Most of the guys came directly from boot camp, so he was older than most, by about a year. That allowed him to have a bigger mouth than most, which came quite naturally.

At night they were expected to study hard, but many "basics" about the aircraft, especially the engine, were already "old hat" to him, giving him plenty of leisure time.

Also, during the day they hung around the flight line, watching other students take off and land, leaving him plenty of time to study and keep up. His evenings were practically free.

On the first Saturday of his stay at Woensdrecht, he rode the train back to The Hague to get his bike and spent the evening dancing with Ria, accompanied by Jeff and Hetty. The clubs and beaches were swinging that time of year. It was mid-summer. A great time to sleep in the dunes or on the beaches.

On Sunday morning he hopped his bike and crossed some of Holland's waterways on his way south. The southern part of the province of ZUID HOLLAND and the entire province of ZEELAND presented nothing but islands, separated by gorgeous rivers, canals and inlets. It was a perfect summer day to be racing through the countryside.

By mid-afternoon he arrived in the ancient market-square of Bergen-op-Zoom, located smack in the center of the old part of town. As in most medieval cities of Europe, the Market Square was completely surrounded by restaurants, bars and nightclubs.

Looking around, "THE CONTINENTAL" caught his eye and he figured he'd check it out. It turned out to be a good choice. The sign at the door read: "THÉ DANSANT" at 4 pm. The price of the beer was reasonable and the atmosphere was cozy.

He was now allowed to wear regular shoes instead of boots and it being summer, they could even shed their jacket and roll up their sleeves. So, compared to the old days, he was well dressed for dancing. The barmaid was pleasantly "chatty" and told the newcomer about their place and other good dance places around the square. The old town seemed like a nice place to be stationed.

At 4 sharp, the band struck up and the place started to fill up a little bit. The problem about checking out a "new joint," no matter where; "*HOW DO YOU KNOW IF A CHICK CAN REALLY DANCE*?" Trial and error maybe.

But what if they play some "easy swing" and the chick can't jitterbug? Or they play a waltz and she doesn't waltz? Pain in the neck! Wait to see her dance with someone else and then make up your mind? You might lose a pretty girl that way.

Back to trial and error.

He hit it right the first time.

A couple of girls had strolled in and found a seat right near the dance floor. One was a striking blonde, the other a brunette, pretty, but not striking.

When the band struck up: *"Oh, It's Only A Paper Moon, Hanging Over A Cardboard Tree,"* he asked the blonde and they "bopped" easily across the floor. She danced differently from the way they danced up north. She didn't easily turn to her left, but it was more like the way they "swung" in Nijmegen, so he adjusted.

They talked a little during the slow dance that followed and she proved to be nice enough, but a little "stand-offish" and cool. After the dance he walked her back to the table, thanked her, and retreated back to his beer at the bar.

When the band struck up again, the blond was asked immediately, so he went and asked the brunette. She was more outgoing with an easy smile, a good dancer, and during a slow dance instead of being stand-offish, she pulled him in closer and they danced cheek to cheek.

After the set he brought her back to her table and found out their names were RODA and BERNIE. While sitting at the bar again he scanned the place and concluded that the brunette was his best shot for the evening.

During the next slow dance he learned that BERNIE was not really a boys name, but an abbreviation of BERNARDINA, while her girlfriends name was actually ROSALINDA and had been called ROSA or ROSIE all her life and had hated it. So, as a teenager, she had changed it to RODA. Besides being different it sounded good he told her, and at the end of the set he transferred his beer from the bar to their table.

They talked and danced till 7, when Bernie had to go home for supper and he rode along with her. He had to be in by 10:30 and there wasn't much point in hanging around town much longer. Bernie lived in the direction of Woensdrecht and on the way, she told all about living with her mother, her sister just got married and moved out, her being a secretary and enjoying it, and before they knew it they were in front of her house. A neat, modest little home with the usual multitude of flowers all over the little front yard.

They stood and talked, bike in hand, about the next time

they'd go out and had just decided on Wednesday when Bernie's mother appeared in the front door. After some introductions she said: "Supper is on the table so come in and get some too. I'm sure you can't get anything to eat on the base at this hour. It's very simple, soup and sandwich, but it's a lot better than nothing."

After a few "I really shouldn't" and a few nudges from Bernie, he found himself in a cozy little kitchen enjoying a hearty soup. It turned out that Bernie's mother had lost her husband to lung cancer, just like he lost his dad last year and she was really interested in knowing how Frits' Mam was coping with the situation.

"Well, Mam has a business to run that doesn't do all that well, but it keeps her mind occupied. Besides, she still has four kids at home: a daughter 18, a son 17 and two more daughters, 14 and 5. So you see, she has plenty to live and work for and she handles it well."

"But when she crawls in bed all by herself, night after night … ???"

"I don't know how she handles that! I guess most times she's just too tired to cry or worry and probably just conks out. Let's at least hope so."

Bernie's Mam felt that he might be right, that being busy was probably the best therapy. "I should go and find a job for myself," she concludes.

"And go out some too." Bernie added: "You can't just sit here and fret."

Everyone agreed and it being 9 o'clock, he started to leave. After a dozen promises to "Mam" that he'd be back, he left.

Bernie would see him off a little ways and this time they're holding hands as they're riding along. Once out of town, they stopped to say good-bye and with their bikes in hand, he tenderly kissed her on the mouth. Tenderly was the way it started, but it heated up rapidly. The kisses got more and more passionate and before long, their bikes were up against a tree and they had both hands free to embrace and explore with. The grassy slope of the ditch along the road was a natural and inviting bed and they just let their lust take its course. She was a fine female specimen of 20 and she loved what a young male was doing to her body. If her watch hadn't warned her that it was after ten and he'd better hurry off, they might have spent a few more hours

Tiger Moth class

enjoying all the aspects of being of opposite sexes, oblivious of passing traffic. When they parted, they changed their date to "TOMORROW". They couldn't wait to get together again.

He made bed check at 10:30 and found out, he only had to be inside the gates at 10:30, so he actually had MINUTES to spare. Nearly every night of the week he met up with Bernie. Rarely, once or twice a week, he'd meet her at her house. The other nights, "MAM" thought Bernie was spending time with Roda or with her sister. Meanwhile, all they did was disappear in the woods and make love.

Even rain was no problem. The military PONCHOS, that button together to make bigger ones and even tents, were solutions to soldiers in WAR, but also in "PIECE."

The only thing that suffered was his studies, because he only had time after 10:30, early in the morning, or on the flight line. Most of the time he was too "pooped" to "pop", and he couldn't keep his eyes open.

One day it hit the fan. Classrooms always had a tendency to put him to sleep, but one morning when "AERO-DYNAMICS" should have had him on the edge of his seat, he fell sound asleep. Major Belderman, a real military minded Indonesian pilot, did not appreciate that one bit. He also didn't buy the excuse that his studies kept him up half the night and had him fetch his parachute. With his parachute banging against his butt he had to jog the entire perimeter of the airfield. The Major needed binoculars to keep track of him at the far end of the field and, not being familiar with Frits' stamina for long distance running, couldn't believe his eyes when even in the furthest nooks of the airfield Frits didn't try to cheat or slow up.

The trip must have been 10 miles, and on that warm summer day it left him so sweated up and exhausted that he considered NOT going to town that night. That thought didn't even last till suppertime. When someone at the table made a joke about "nipples" he knew he had to go. And he did!

The flying part was intriguing to him.

Their job at Woensdrecht was to qualify for further training. The schedule was as follows: 25 hours on the TIGER-MOTH, completed successfully, got them off to America. A little less successful, but still qualifying, would get the young pilot transferred to GILZE-RIJEN, for training on the AT-6, HARVARD, with a possible shot at the GLOSTER-METEOR.

Most of them WASHED-OUT. Less than 45% of the young men that started in Woensdrecht would make it through the 25 hour course. Dual instruction was standard through the first 12 hours, at which time the "all important 12 hour flight check" was flown with the "almighty flight commander!"

That was the point when most "wash-outs" were dishonored by being sent home. Some of them never made it beyond the first flight. Some would faint, throw up or beg to be brought down. Most of the "wash-outs" simply never solo-ed. Frits didn't understand the difference between the guys that "had it" and those who didn't. One day he'd be looking up to a guy who'd stand there and brag about how perfectly he "rolled" her, how he "spun and recuperated" beautifully and whose landings were all "eggs", and the next day that same guy would be OUT and on the way home.

It didn't seem all that complicated. In class they were taught the theory: why did the plane "FLY"? Why "ROLL?", "YAW?", "SPIN?", "GLIDE?", "SLIP?", "CLIMB?" and "STALL?" All of it made sense. Then they got into the plane and put it into practice.

Tiger Moth

If he wanted to go up, give it the throttle. Go down? Pull it back. If he wanted to turn or bank the wing, just push the stick to that side and push the rudder in the same direction until satisfied with the turn. Keep the nose from sliding down, and that was all.

That seemed simple enough. Making loops and immelmans were fun. He just had to make sure he didn't drift in one direction or another so he would come out of the loop heading for the same point as when he started it. Spinning was easy enough. The plane easily dropped into a spin and easily came out of it.

The main trick was landing. That's where most people failed. The "doghouses" outlined the landing strip and the ground crew would move them as the wind direction changed. That way, they always landed directly into the wind. The reason why those wooden crates were called "doghouses" was because that's what they looked like.

Coming down on the "FINAL" leg of the landing with just the right airspeed and the throttle closed would result in a smooth "flare-out," and by pulling back on the stick, back, back, all the way back as the speed reduced, would result in a "THREE-POINT" landing or an "EGG!" The main wheels and the little metal slide in back would hit the ground at the same time and stay there. NO BOUNCES! That was the ticket! Some did it, some did not.

At 10 hours and 10 minutes of flying time, chubby Lieutenant Hoxel climbed out of the front seat, fastened the straps and said: "Now remember, a lot of weight just got out of that plane and she'll be OFF in seconds. She'll also float longer, so when you come in for landing, pull your stick back more slowly so she won't climb on you. Just do the same thing though. Back, back, back all the way into your stomach and you'll be alright. Now show me three "EGGS" while I watch from here."

Off Frits went, zigzagging his way to the end of the runway. When everything checked "clear", he shoved some power to it, lined up between the doghouses and gave it the gun, all the way forward. Hoxel was right. His tail came up in seconds and he was off the ground before he knew it. Without the 220 pounds in front she climbed like a fighter.

Up in a plane, all by himself, was the absolute ultimate sensation of his life. It even beat sex. At 400 feet he started his right turn and feeling as free as a bird, he started to sing at the top of his lungs. This was the only way to live!!!

Turn right onto "down-wind" at 800 feet, parallel the runway under your right wing, *LALALALALA*, what a way to go!! *LALALALALA!* When parallel with the last doghouse he cut the throttle, set up his glide, at 600' right turn onto "base" leg. He's talking to himself now, another 90 degree turn at 300', line up with the runway, size up your height, good! Cut back the throttle all the way. Remember you're lighter now. She'll float longer. Don't give more power, you'll make it Fritsie-baby, ya ya, back, back slowly, back, back and ... bump. Down on all three. BEAUTIFUL!

"Give it the gas, stick forward, up comes that tail, back a little ... we're flying*! LALALALALA, LALALALALA*. I should have started this years ago!"

Again singing all around the pattern until it was time to prep for landing and then he started talking himself through the procedure with the same results, and after doing it one more time he taxied back to the instructor. He watched "chubby" climb back in and took him to the ramp, some half-mile away.

Lt. Hoxel proudly signed off his "pre-solo" flight in his logbook and MOST proudly Frits filled in his first SOLO flight and signed it off. That was something to celebrate. And he did!

THEY did. His buddies tried to get him drunk but he escaped in time to celebrate with Bernie. She was as enthusiastic as he was and they both felt as high as he had felt today when he was up there. It was a grand day in his life.

The following morning the Lieutenant told him: "Today, we're gonna run through everything that you're supposed to know up to this point, 10 times if necessary. By the time we get back you'll have just about 12 hours and you'll take your test with Major DeKoning at 11 and you'll pass it! So let's go, you got your work cut out!"

And they did. Climbing turns, shallow and steep, gliding turns, same menu. Steep level turns, inverted flight, slow roll, spin, loop, immelman and three landings.

At 11 sharp, the dreaded Major showed his affable side. He gave subtle orders: "Do this, do that", made no comments, asked for their position twice and after one landing told him: "Take me back."

Frits' heart stopped!!!! He thought he did pretty well. "*WHY SUCH A SHORT FLIGHT WITH SUCH AN ABRUPT END*?" He didn't understand!!!!!

After they parked the plane the only order he received was: "*BRING YOUR LOGBOOK*!" And the Major disappeared into his office.

Lt. Hoxel walked up, "How'd it go?"

"I don't know, I think I did alright but he acts so weird!"

"What did he say?"

"Bring your logbook."

"How long were you gone?"

"20, maybe 25 minutes"

"How many landings?"

"Just one, that's all. That's what ha... "

"Well", the instructor interrupts, "Was it a good one?"

"Yeah, standard, three point, no bounce, same thing as this morning, right between the first and the second doghouse. I don'... .

"Did you have to use power?"

"No, once I cut it, I left it off. I didn't need any.:

"You're alright, let's go see him."

However, the Sergeant-secretary asked them to wait. "He

has someone in there."

Frits meanwhile logged in the plane #, the date, 12 hour-test and the things they did, maneuvers performed. Everything except the most important; the signature from the test pilot.

After a while they were asked in, saluted and stood at attention.

"At ease, gentlemen!"

Major Belderman was present too, leaning against the window sill. The abominable Test Pilot sat behind his desk writing something. Cold sweat was running down Frits' back. Somehow he felt his young flight career had just come to an abrupt end.

"Frits?" The Major used his first name. WOW! "How do your parents feel about your flying?"

"Well, I have no father anymore and my mother, well, she's like a woman." Chuckles all around. "She says it's alright, but why do I wanna be a jet fighter pilot and kill myself, like Karel Kohler, a friend of my .."

"You knew Karel?" the Major interrupted.

"Yes, my brother dated his younger sister for years, so I told her I'll never die in an airplane. I'll die in bed with a wild woman!" More chuckles. "And she said, I'm fresh and should be ashamed of myself. Hahaha, she might be right!" He'd forgotten all about his nervousness.

"How would she feel about you going to America?"

"Sir, I've been ready!" "Sir" he added fast.

"Lieutenant, you need to teach him crosswind landings, all the stalls, anything you haven't covered yet and send him solo the rest of the time. Fly his butt off. He needs to complete his 25 hours in a hurry, rush on home, get his civvies and meet the boat in the middle of the month. He's complementing the preceding class!"

"Yes Sir, Yes Sir," Hoxel's grinning from ear to ear.

"Frits let me sign that book and shake your hand. Good flight, boy, good flight!"

"Thank you Sir."

"Thank you!" The other major also shook his hand.

"Dismissed."

Once outside the door, Hoxel gives 'm a hug. "Go eat, you rascal. I knew you could do it! Check back with me after chow and we'll work up a schedule and at 5, I'm buying the beer!"

"HALLELUIAH"

In the mess hall where his buddies were already eating, his enthusiasm flared up again. *"HALLELUIAH, I'M LEAVING ON THE NEXT BOAT FOR AMERICA! THEY'RE GOING TO FLY MY ASS OFF. "THAT'S A QUOTE!" I have to finish 25 hours in a hell of a hurry but it will be fun!"*

His classmates congratulated him and he was too carried away to notice if there was some envy or not. It didn't matter. They knew that he had solo-ed at ten hours yesterday and had his 12 hour ride today, so he had to be doing something right.

He got one more flight in that day, mostly crosswind landings, or rather introduction to crosswind landings, because there was very little wind that day. Also power-on stalls, straight, left and right turns, full power stalls, like on take-off, partial power, like cruising and power-off, like coming in for landing. It brought his total hours up to 13:40, it had been a full flying day, but boy, had it been exciting!

Over a beer with Lt. Hoxel he learned that the jolly instructor knew he "HAD IT" from the second flight.

"Some people can fly, some can't. The ones that can't aren't losers, never look at it that way, they should just go into a different business. It's like this: Just because you and I can't paint like REMBRANDT, we're not losers, but we should have enough sense to stay out of the painting business, right?"

"I'll drink to that, no painting business for me*! PROOST*!"

Thank God, after a few more beers, the Lieutenant had to go home to eat, so Frits sped to town and got an unexpected reaction. Bernie stormed out of the kitchen, shouting: "I hate you, I hate you!" Mam said: "Have some more soup, she'll get over it!"

Mam was right. After a while, Bernie rejoined them and even got a little excited when he relived the flight and the ensuing meeting. Once in the woods, she cried: "You'll forget all about me, I know, I know. With all those good-looking girls in America you just won't think about me, I know, I know!"

She put a damper on his great mood. He made love because he was expected to and excused himself early so he could grab one more beer at the bar, near the gate. This was his great day and NO WOMAN was going to ruin it. Thank God for beer. He slept like a log.

Flying every time a plane was available was fun. Especially showing off in front of the other fly-boys, hanging around for their next ride.

Twice more he flew with Hoxel and the crosswind landings were exciting as hell. That man Hoxel could land that little bi-plane on one wheel, wing dipped into the wind, add a little power and ride the whole length of the runway on one wheel before gunning it for take-off. He'd cross-controlled it like crazy, his stick in one direction, his rudder in the opposite. Frits ended up with a certain goal; run the length of the runway on one wheel before his 25 hours would be up.

Well, the 25 hours went by so fast that he never did get to do his "WHEELIE." Before he knew it he checked out of Woensdrecht, complete with new uniforms, including three khaki ones, a trench coat and travel papers. He was given a bunch of shots and was off for home.

The farewell of Bernie consisted of a brief stop by her office on the way to the railroad station. That was it. No time for tears or sentimentality. Life was flying by.

The train ride gave him time to reflect on all the things that had happened to him in the last few weeks, but most of all on what lay ahead.

America was something you read about, some faraway dreamland that you see in the movies, yet now he was on his way there. Just a short stop at home, one brief night, just enough time to pack some stuff and say goodbye. What a mind-boggling thought.

At home, everybody was enthused about his upcoming trip, but he couldn't tell anybody where he was going in the U.S., because he didn't know. All he knew was that they were docking in NEW YORK.

He turned down all offers to see him off to the boat in ROTTERDAM, and by 8:30 on a sunny July morning he boarded the bus to ALKMAAR in order to catch the train to Rotterdam where, in turn, he'd catch a boat to New York and a tremendous adventure. After waving goodbye to everyone in front of the little CAFÉ RUSTWAT, he glanced at his watch that Mam just gave him. His first watch in his life!

CHAPTER IV
AMERICA!

The ship looked tremendous. It wasn't really, but to him it was tremendous.

The "ZUIDERKRUIS" (Southern Cross) had been an ocean liner, a troopcarrier during the war and after that, mainly hauling Dutch Emigrants to Canada. The result was that there was a sizeable Emigrant section with six bunks to a cabin, and that's where the Dutch Fly-boys ended up.

He found his cabin, met some of the boys and after shaking hands all around, changed into civvies for the first time in a year and a half, and headed back on deck to wave off the crowds that stayed behind.

Hundreds of people lined the railing, waving at hundreds more that lined the dock. It seemed that everyone else aboard was being seen off by throngs of family and friends and, to his surprise, lots of tears were being shed on the ship, as well as on the quay. He couldn't understand that. This was a moment of momentous joy, an occasion to celebrate. Certainly no time for weeping!

After the ship shoved off he was informed that he was expected on the rear deck in uniform. It was one of the guys he had met in the cabin.

"What for?"

"The Lieutenant called a meeting and he wants everyone

there!"

"A Lieutenant? I thought we were all just privates."

"Oh, no," WIM explained. "We have two First Lieutenants, one Sergeant, one Corporal and 31 Privates."

"What's all that brass for? Can't we get across an ocean by ourselves?"

"They're also Student Pilots, that's why!"

Frits couldn't believe it. "We have N.C.O.'s and Officers as student pilots?"

"Yep, and the "LITTLE SHIT" has seniority, so he's in charge, so you better come."

"Piss on him! I'm on an ocean crossing on a civilian ship and I like it that way. I'm remaining a civilian. You don't know me and you haven't seen me. Never heard of me, O.K.?"

From inside a lounge area where small groups were playing cards, he could look down on the lower deck without being seen. He watched the Dutch Fly-boys lining up while two Officers shouted orders. The LITTLE one did most of the talking and inasmuch as he couldn't hear a word that being said, he figured it was time to check out the BARS!

He located the bar without a problem but received some good news and some bad news. The good news was: a glass of "HEINEKEN" beer was only 20 cents. His money would go a long way. The bad news was: the bar wouldn't open until they were 3 miles out to sea and right then, they were still on the NIEUWE WATERWEG on their way to the North Sea.

Oh well, might as well go check out the scenery and the ship. NOT the stern. He climbed his way to the upper deck on the bow and watched the famous "SCHIEDAM JENEVER" and the "SHELL REFINERIES" floating by. Rather, THEY stood still, he floated by. He had to get used to this.

The weather was exceptional and the slight breeze, caused by the forward motion of the ship, took the edge off the heat and made it downright pleasant. While eyeing the people on the decks, hundreds of them, he contemplated possible partners, singing, dancing or bed partners. There was some good potential amongst them.

Most of the people aboard were emigrant families though, Pap, Mam and a few or a bunch of kids, seeking better futures for themselves.

He was rudely disturbed by a voice that reminded him of JOSEPH "YUPIE" GÖBBELS: *"SOLDAAT FORRER, I BELIEVE?"*

"YES SIR!" He knew before he turned around it had to be the *"LITTLE SHIT"*, and it was.

"Didn't you know you were to report to me immediately upon boarding?"

"Who are you?"

"I'm Lieutenant Harding and I'm in charge of this DETACHMENT."

"Well sorry Lieutenant, I've never heard of you and I don't belong to your detachment."

"You don't, do you?" The little man really looked annoyed.

"No, I finished up in Woensdrecht yesterday and I'm joining up in the U.S. with the class ahead of me."

"You finished up yesterday?"

"Yes, I flew my 25th hour yesterday morning and here I am. But it was nice meeting you, Sir."

"Well, Private, whatever. This is the class you're joining, right here and ... "

"Where?"

"The class that's on board here, MY detachment of which you're now a part and I will not allow civilian clothes while on a military move, so go to your cabin and change and report back to me."

"For real?"

"You don't believe me?

"COME ON! This is a cruise, we're here to enjoy it!"

"Soldier, I can still place you under arrest and have you returned by the pilot-boat."

"Sir, I just left. Sir!"

With some sort of salute he disappeared.

"That miserable little runt! He might ruin the whole damn trip."

Cursing under his breath he made his way to the cabin where one of the guys, "JAN", explained that the RUNT always had that kind of attitude, while the other Lieutenant "JERRY", was more like one of the boys.

"Let's arrange to throw him overboard!" was all Frits could think of. Then he shrugged; he'd find some way to make this trip fun anyway. When he encountered a ship's Officer, he stopped him and asked: "Sir, does the presence of military men in uniform have a depressing effect on the Emigrants?"

"Well … I don't know … " The officer was a little baffled.

"It's very reminiscent of CONCENTRATION CAMPS and OCCUPATION, isn't it?"

"Well … "

"You ought to tell your Captain that no military uniforms or formations should be allowed during the trip, for fear of upsetting the passengers."

By now the man was smiling. "I sure will tell the Captain, Sir."

"Thanks."

He could have saved his breath. Lt. Harding got seasick, the moment the boat hit open water and he was hardly seen for the duration of the trip. So much for that problem.

The other Lieutenant, whom Frits met in the "LOWER" bar, with bartender "OME KEES", was of a different breed. While Ome Kees poured beer, the Lieutenant introduced himself as "JERRY" and stated his opinion of the situation. "As far as I'm concerned, we're all student pilots with just 25 hours under our belts and rank doesn't make anybody a better pilot, so until I beat you in the air and prove myself to be better, we're at least kinda equal and maybe you're even better."

"I'll drink to that! "*PROOST!*"

When Frits told him what he had told the Ship's Officer, Jerry laughed uproariously. "I'm gonna tell Harding the same thing. I'll say that that's what was filtering down from the Captain, that people were getting edgy and all military display should be discouraged. *HAHAHAHAHA*." He roared! He thought that was so funny. "And when Harding asks me, who filtered down that information, I'll tell him you did. *HAHAHAHA*! Kees, give that man a beer!"

"Thanks, Happy flying."

"Hey, Frits, this didn't come from me, but spread the rumor that civvies are really allowed, even preferred! O.K.?"

"*O.K.! PROOST*! This is gonna be a fun trip!"

Jerry explained how he'd been drafted in the Army, then sent to O.C.S. One day, as an Officer, he rode in an "AUSTER", a light British built plane, as a FORWARD ARTILLERY CONTROLLER and he fell in love with flying. So he asked for pilot training, was tested, passed, and joined the Air Force. In a way he was even newer to the Air Force than Frits was, which was another reason to order a round of beer. *"PROOST!"*

At dinnertime half the guys showed up in civvies and Lt. Harding didn't appear. Because the ship's management had placed all the aspiring pilots at the same sitting, in the same area, Frits got to meet most of his new classmates for the first time. It seemed like a jolly bunch of boys, from many varied backgrounds and ideals. Some seemed very serious about this flying business, and some of them thought of it as one big happy ball. The latter group appealed to Frits.

When the band struck up, he entertained some older ladies, his mother's age, by asking them for a waltz, a foxtrot or a polka. There weren't that many young folks aboard but a group of girls from SANTURCE, PUERTO RICO, certainly had a way with the Rumba, the Tango and the Samba. One of them, a little thing barely 5 foot tall, was quite a jitterbugger as well.

Unfortunately, most people retired early and only in the "DOWN BAR", a little nightclub way down in the hull, there was some action left after 12. The piano player kept the atmosphere most pleasant and before long he had some amateur singers "WORKING" with him.

"There's Just One Place For Me, Near You. It's Like Heaven To Be Near You!", A Trio was swinging away in no time. Frits played the bongo drums and another fly-boy, Jan de Ruiter, was doing the vocals.

"Into Each Life, Some Rain Must Fall, But Too Much Is Falling In Mine." "Sing it ... Ella!" The whole little bar was bussing along. On the dime-sized dance floor people got into the act.

For the Canadians aboard: *"J'attendrai, Le Jour Et La Nuit, J'attendrai Toujours!, Ton Retour!"* He had Jan on this one. Jan knew every American and English song that Frits knew, but French was not his forté.

The music and the beer flowed well into the night and like all good things, it had to end. What a day though. *WHAT A DAY!*

WHAT A DAY, the next day and the next! For the passengers aboard the liner, every day was a glorious day.

The "Ping-Pong" tournaments, the "spoon-diving" contests that earned him a bottle of champagne, and another bottle in a "Jitterbug" contest. It was all so grand ... trapshooting off the stern, lounging in the lounge, watching two SANTURCE sisters play the living hell out of the piano with their short little fingers, or just lounging around the pool. It was all one big long holiday!

The Puerto Rican girls, Carmen and Judy, (HOODY), especially intrigued him. When the little one, HOODY, told him she worked for the F.B.I. she was astounded that he didn't know what that was and that he had never heard of it. She assumed that the whole world knew and feared the F.B.I. She had studied at the University of Puerto Rico and at Georgetown in Washington, and both her English and Spanish was flawless. What was more important, those girls had rhythm, in their fingers and in their feet. Most astounding!

The two of them could sit at that piano and whip up the fastest sambas and mambos he ever heard. They could dance to these beats with equal ability and also sway to a Rumba with such grace that they soon were the undisputed "BELLES OF THE BALL."

No one really knew where they would end up in the United States, but Jan de Ruiter and Frits were convinced that they'd end up on the same base somewhere and tear up the town as well as the skies.

It didn't work out that way, but on board, it made for a happy fantasy. The food was excellent, the weather was great and the ship pulled into and out of SOUTHAMPTON, ENGLAND and CORK, IRELAND to pick up additional passengers, and that made the group even more cosmopolitan.

The 5th day out, they hit a rare summer storm and everything on deck was tied down. Ropes were tied all around so the courageous ones had some sort of a hand-hold when they ventured outside. It was exciting. The bow would dive into the waves and the propellers in the stern would roar as they came up out of the water. Thank God and the seamen for the ropes, because the bucking ship would surely have thrown a few overboard otherwise.

Dancing was impossible. The dance floor pitched too much and not too many people showed anyway. The same with chow. Soup couldn't be served. It would have sloshed out of the plates and with so few people showing up for meals, the few that did got all the attention. Talk about getting spoiled. The finest of the finest that the ship had to offer, and loads of it. Frits had never eaten so well or so much in his life. Three lobsters one night, two Beef-Wellington's another night, three dozen shrimp for lunch. It was all too good to be true.

The small group in the "DOWN-BAR" at night had a great time, but when it was time to hit the sack at about 4 in the morning, he was in and out of the cabin in just 20 seconds. Four of the guys were in their bunks groaning and moaning, but to make it worse, the place stunk to high heaven. They must have been throwing up all day and all night. So he grabbed his blanket and took it to the upper lounge where he bedded down on one of the benches, clothes and all.

The money might not have held out, but the trip could have gone on forever as far as he was concerned. Good weather or storms, it didn't matter much. Everything about the trip was exciting to him.

There were some interesting characters amongst his buddy pilots and some he could easily detest. One of the guys he palled around with was another FRITS; Frits Flieger. He was funny, gentle and very sentimental. Every time he got a little loaded, he fell in love FOR LIFE. If he didn't get to escort his new love to bed, he'd cry big tears of rejection. He was funny that way. His sadness would last at least until the next pair of good-looking legs walked by.

All the guys in the group spoke English. It was compulsory in school in Holland, but Jan and Frits spoke American. That was probably because of the hundreds of American songs they knew and the hundreds of American movies they had seen. Besides, they were getting plenty of practice on board.

There had not been even ONE more meeting during the whole trip, but on August 4, 1952, word reached everybody: "Uniforms tomorrow! When docked, meet with hand luggage in the AFT lounge, ready to travel!"

Long before America came into view they were lining the decks looking for land. When the sun came up, a slim, gray line on the horizon was the first sign of land. That line grew as they sailed closer, until that line became Coney Island. Anybody who had ever seen a movie about New York recognized the Parachute tower and the Roller Coaster rising above that gray line.

As they sailed up the harbor, they passed Miss Liberty in all her glory with the famous Manhattan skyline behind her. "Strange," Frits thought to himself. "It all looks so familiar!"

They had seen those scenes dozens of times in the movies or newsreels, and it made it seem like they'd been there before.

The Brooklyn and Manhattan bridges appeared briefly and slid out of view as the ZUIDERKRUIS moved slowly up the Hudson River.

All of a sudden, America lost all of its luster, its magnificence. The factories, warehouses and docks looked ramshackle and weather-beaten and old. The houses of Jersey City and Hoboken looked very old and dilapidated. They had never seen anything so depressing. In Holland where everything was nice, tidy and freshly painted, they had never thought of a run-down scene like this.

It got worse. Disembarking onto a blue Air Force bus that was clean enough, but looked like something "PRE-WAR," they sat at eye level with the grimy bars that lined the Hoboken waterfront. Trash in the potholed street, dirty windows and sidewalks, rusted, dented big old cars ... it seemed all so unreal, so un-American. It had them stunned. The people, mostly Spanish looking, were unkempt and the kids were downright dirty.

After crossing through the Holland tunnel, a cry rose from their mouths and minds: "Is this the famous "Canal Street?"

"Is this really America? This is nothing but a dump!"

Even the few neon signs of Chinatown did not add any luster. As a matter of fact it highlighted the contrasting rubble. They were terribly disappointed.

Thank God, the Manhattan bridge brought a different kind of scene into view, and from the top of the span they could see Miss Liberty again on one side and the Brooklyn Navy Yard on the other.

Their destination: Fort Hamilton in Brooklyn wasn't bad, and there is where the Lieutenants were issued papers for all of them, after they were accounted for.

Back on the bus. This time to the St. George Hotel in Brooklyn where good rooms and a nice view awaited them.

A brief meeting was held where everyone received their room assignments and meal tickets. And ... their travel orders!

"TRAVEL ORDERS? Where are you going? And You? And how about you?". Oh, DAMN. Jan was going to BARTOW, FLORIDA, Frits Flieger to someplace in Northern TEXAS, and our Frits? "KINSTON, NORTH CAROLINA!"

"CAROLINA, what's in CAROLINA?"

"Nothing good, that I know of."

"Who else is going to North Carolina?"

"Me!, Here!" Louis Borst, known as "LOOTJE", a short red-head with a funny face. "HI, Loutje! WHO ELSE FOR NORTH CAROLINA?"

"WIM?" "Win van Oosterbeek." "HI, Wim."

"WHO ELSE? NORTH CAROLINA? ANYBODY?"

"Where are you going?"

"Greenville, Mississippi. And you?""And you? And you? And where are you going?"

"When are we leaving? Let me see the ticket. Pennsylvania Station, Monday 8-7-'52, 11 AM."

"Hurrah, the whole weekend in New York! New York, New York, It's A Hell Of A Town, The Bronx Is Up, The Battery's Down."

"Hey guys, where do we go first?"

"TIMES SQUARE! BROADWAY!"

"RIGHT!!! Come On Along And Listen To, The Lullaby Of Broadway. The Hip-Hooray And Bally-Hoo! LET'S GO!!! Are you ready??? Let's go. Let's hit it!"

"WAIT, WAIT, SOLDIERS ... WAIT?"

"What the hell does the RUNT want now? Wish he had stayed sick."

"Make sure that you are here in plenty of time to catch the Air Force bus to take you to your respective trains."

"Don't worry Lieutenant." To himself he murmured: *"Drop dead!"*

As soon as they were in their civvies they were off. Down into the subway to Times Square.

"*What a place*!" A bit messier than they expected but still grand. The only problem really, was the heat. It was oppressive. Most of them had never experienced heat like that before. The difference in temperatures between the soda fountains or stores and the outside heat hit them with a bang every time they stepped back into it.

Most of them had never been in any kind of air-conditioned room until they boarded the ship, but on the ship it never got this hot.

Of course they bought all the 'NECESSITIES" that every visitor to New York buys: a PERRY COMO shirt, a miniature EMPIRE STATE BUILDING, a "GOLD" plated I.D. bracelet for a dollar, 25 cent pictures from a booth and 15 cent Pizza slices.

Walking up Broadway and staring at every marquee was very important. Watching a giant John Wayne puffing smoke out of a Camel sign was as impressive as Miss Liberty had been. Frits HAD to see 52nd Street. The radio program in Holland: "SWING AND SWEET FROM HOLLYWOOD AND 52nd STREET" had been a favorite of his for years, so he just had to check it out. On a hot Saturday afternoon there wasn't anything much going on in the clubs, but the one they couldn't resist was one off 7th Avenue on 51st, called "BIRDLAND". Charlie Parker was the featured entertainer and, of course, he wasn't there at that hour, but his picture was plastered all over the place ... his and dozens of other Jazz Virtuosos.

The place was downstairs, below a storefront and was dark and cool and nearly empty. They talked with the bar-man and the few customers and learned the obvious: this place and the other clubs didn't liven up until after dark, so after just one expensive beer they headed out. It was too early for the "before theater" crowd and the real crowd was not expected till after the theaters let out, after 11.

When Frits had asked about visiting HARLEM the customers in BIRDLAND strongly advised against it. He explained that his Dad was born in HAARLEM, in Holland and that he wanted to see the place in the worst way, but their advice remained the same: Too dangerous, stay away from there. His buddies wouldn't go after that advice, so that ended that.

Before getting back into the heat they had agreed to head back to the St. George for chow to save money that way, and then head on to Coney Island. So they sauntered down "The Avenue of the Americas" and boarded the subway at 42nd street. The subway system seemed complicated at first, but once they understood the green, red and blue lines, it became more simple.

Their image of America was formed by magazines and movies, and that image was severely tarnished again by the subway system. Some of the guys had been in London's UNDERGROUND and Frits had been in the METRO in Paris, and in comparison, the New York subways were a pigpen. The important thing though was, the dirty, shaking, clanging subways got you to where you wanted to go, and right then it brought them back to the St. George.

Their first American meal was good. They had looked forward to "T-BONE" steaks, but the menu didn't offer that in

Brooklyn, but the New York Strip was excellent. If only they had enough money they could live like this forever.

The heat subsided a little bit and by the time the "SUBWAY" became the "EL" on the way to Coney Island, it even became bearable, as long as the train moved.

Once at Coney Island it became America again as they had envisioned it. Their eyes just popped. None of them had ever seen so much neon and glitter in their whole lives. It was utterly fantastic. The height of the roller coaster was most impressive and it's ride was more scary and gut-wrenching than that first ride in the TIGER-MOTH not too long ago.

For the first time in their lives, they ate corn. "Corn on the cob, 5 c," lured them over.

"In Holland we feed this to the pigs!" was their comment, but most of them bought one anyway. With some butter and salt Frits found it kinda appetizing, but some of the guys took just one bite and threw it away, never to try it again in their lives. "You guys probably never went hungry!" was Frits' comment "Otherwise you wouldn't throw away good food!" They didn't care.

The one disappointing thing about the evening was they didn't meet any girls. There were plenty of them around and some nice looking ones at that, but it seemed impossible to get to talk to them. When they asked some of the stand operators where they could go and dance they were told: "Times Square, dime a dance!"

Well, there was no way they'd hike all the way back to Times Square and after hardly having slept the night before, they dragged their tired butts back to the hotel at 12:30.

In the St. George bar, a neat little combo was playing some good dance music, but there wasn't a single girl in the joint. All couples ... so after just one beer Frits joined his buddies upstairs and hit the sack.

The following day was exciting, but exhausting. Besides that, they timed it wrong. They should have ended their day in "Greenwich Village", they probably would have had some fun there. But early on Sunday morning there was nothing much exciting about the "Village" and although it was interesting, it was quiet and dull.

They walked all the way to 34th street, through little Greece and did find the Empire State Building and the view magnificent and exhilarating. The heat made the view a little hazy but there was enough left to see to make a lasting impression on their minds.

The walk to Rockefeller Center was stupid. They should have taken a bus or the subway. The heat was drenching and there wasn't much of anything going on in Rockefeller Center anyway. A bunch of the guys wanted to see SAINT PATRICKS, but Frits got on a bus that took him up 7th Avenue through the middle of Harlem.

It was hot, but most fascinating. Having been raised in a country where, for the lack of Negroes, at St. Nicholas time, they'd paint people black to work as the Good Saint's helpers and, never having seen a Negro until he was 12 years old, he certainly got an eyeful here. Once the bus turned north from 110th Street he didn't see a white face, unless it was a cop. The Sunday heat must have driven them into the streets, because it was crowded with black faces. Music blared out of open doors and windows and generally, everyone seemed to be having a good time. There was a lot of laughter and happiness all around and he was tempted to get off and buy a beer.

He didn't though. He just rode and took in the scenery. "Sugar-Ray's" lounge surprised him. He thought of him as a fighter, not as a nightclub owner. Looking to his left, as they crossed 125th street he spotted the "PALM CAFÉ " and the "APOLLO THEATER". He had heard of those. A lot of musicians had gotten their starts there and it made him think of "*JUMPIN' WITH SYMPHONY SID*!" "*OOBEOO POP. BEEBEEOOBAH*!" Boy, he'd love to jump off and explore this further. Even though he really didn't feel scared, the warnings about Harlem were still ringing in his ears.

North of 125th Street he could see the old BROWNSTONES along the side streets with their front stoops, reminding him that Harlem had been a classy "uptown" neighborhood at one time. Turning left on 135th Street the number of bars started thinning out somewhat, and after a few blocks they started climbing into a more residential neighborhood.

Again the bus turned north, moved onto an avenue lined with stores and the brown faces became interspersed with Latin and Caucasian ones. At 155th Street he got off and walked into "TONY'S TOWN". It was dark and cool in there and before his eyes even adjusted to the darkness he grabbed a stool and ordered a beer.

"Wha Kine?", the dark face asked.

"Balantine" was his answer. He had tried several brands and he liked that one.

"We ain't got no Ballantine, we got Schaeffer, Rheingold, Bud, Pabst an' Miller."

"Rheingold sounds good." He had seen "MISS RHEINGOLD" pictures on the subway. If the beer tasted as good as the girls looked it should be great. It was O.K. It was good and cold and that was the important thing.

"Fifty cent," the dark face ordered.

He had to get used to that, putting money on the bar first. Everybody in the bar had money in front of him, so that must be the way they did things here.

Fifty cents for a bottle of beer was expensive so maybe they had entertainment in back he figured, so he picked up his bottle and strolled to the rear. The bar was very long. 40 barstools long, at least. The end widened out into a seating area with a jukebox, but no entertainment. And not a white face in the crowd.

The people seemed friendly and certainly nothing to be scared about. He made his way back to his stool as the air-conditioning started to dry out his sweaty clothes, and continued enjoying the ice cold beer.

"You a cop?" The black face behind the bar was more visible now.

"Huh?"

"You a cop, man?"

"A cop? A policeman? No, I'm not. I'm from Holland."

"You'se who?"

"I'm from Holland. The Netherlands ... Europe."

"Whea in Europe?" The man next to him got into the conversation. He was really black, whereas the barmaid was more Mulatto, more yellow.

"Holland", he repeated, "Amsterdam."

"Oh, you is from Hawland? I bin dere, in 'Fowty-fow', I drove trucks all over EUR'P. I was in Mostrik."

"Maastricht, in the southern part of Holland."

"Yea, man, Mostrik, Ahken, Cologne, all ova dem pahts. Naish country. Josie, giv dis man a beah! Dem people was good to me, dey gave me IGGS, lotsa IGGS. We done had nuttin but powda IGGS, but dem Hawland people, dey give me lotsa IGGS."

"That's great. I lived further north. We didn't get liberated till April of '45, after that nasty winter of '44-'45".

"Thanks" . The beer arrived and he chug-a-lugged the remainder of his first bottle.

"Yeah, man, I remember dat winda, I drove gas trucks through the Ardennes Mountains in da sno', trying to keep up with Patton an' all his tanks, yea man, dat wao won hill of sonofabitchin winda. I member! Y'all wanna eat sumpin, man? Who's you name, man?"

"Frits."

"Freetz? Da's German!"

"No, in Holland we spell Frits with an S, in Germany they spell it with a Z."

"Whao's you doin' hea, man?"

"I'm on my way to North Carolina for training, I'm in the Dutch Air Force."

"You is? You a pilot?"

"Not yet, right now I'm still a private, it'll take a year or more before I become a pilot."

"I was a Sargint in de Army."

"Should I salute you, Sir?"

"No, no, no, no, no. 'daint necessary. Josie you got some of dem lobster-tails left? Bring me some."

"How miny you wants, Willy? Free-foa?"

"Fo!" Sergeant Willy explained, they had their usual Sunday special. Lobster-tails, little ones, with bread and butter. Just $2 a piece and that's why so many people came out on Sunday afternoon. Frits liked the way he pronounced "SUNDI". Willy insisted that Frits have two of the "tails" and half a loaf of CUBAN bread and loads of melted butter. It was delicious.

After his fourth beer, Frits made up the story that he had to be in the St. George by 7 o'clock and with a lot of protesting and insisting that he couldn't pay a cent, they finally waved him off.

The heat hit him smack in the face as if he walked right into a wall. The folks had told him to catch the subway right out front, but change at 42nd street to the Canary or Canarsie line or something.

When he woke up he was on the EL, overlooking private homes. By studying the map he concluded that he might be in

Queens somewhere. His navigation courses in Woensdrecht and the blue line on the subway map told him so.

He got off at the next stop, soaking wet and decided "ELMHURST" seemed like a good place for a cold beer. His throat was so dry he thought he swallowed the entire SAHARA desert. Down long metal stairs he landed under the "EL", that rattled and shook with every passing train. The noise was something else. He couldn't even hear himself think.

A "RHEINGOLD" sign blinked at him, telling him telling him *"COME HERE. THE BEER IS NICE AND COLD."*

He crossed over to the "SHAMROCK" on the other side of the street and gasped at the cold air inside. When his eyes got adjusted again he saw nothing but white faces and not a stool available. He positioned himself at the end of the bar, plunked down a dollar and ordered Rheingold. He received a glass full and 90 cents change. Now that was better. After a trip to the filthiest bathroom he'd been in since France, and a few more beers, he hiked his way back up to the EL platform and this time made his way back to the St. George.

Everyone was gone so he ate by himself, showered and took in the music in the bar. By nine, he started nodding in his chair and he figured: enough is enough.

In the morning at Penn Station, the good-byes that had started at 7:30 when the first group left, finally were concluded when eight of them boarded the train for Washington, D.C.

On the bus on the way over, much to the chagrin of the little RUNT, Jan and Frits gave a "MODERNAIRES" version of the *"Chattanooga Choo Choo!" "Track Twenty-Nine, Boy You Can Give Me A Shine!"* And that led to a rendition of *"Have You Ever Passed The Corner Of Fourth And Grand? Where That Little Ball Of Rhythm Has A Shoeshine Stand?"* Even the runt got in the finger-snapping a minute when they hit: *"With That Hoppity, Hippety, Hoppity Hippety, Hoppity Hop!"*

The bus was swinging. The negro Airman who was the driver swayed and rocked right along with the beat of the finger-snapping, but it was over too soon and they were off to their designated tracks.

The trip was interesting some of the time, but the landscape was mostly flat, like in Holland. In D.C. the three boys, headed

Taxi-ing in Preflight

for North Carolina, got off and waved the other Fly-boys farewell as they pulled out for the trip out west.

They had a 48 minute lay-over and thought of going into D.C. but the train personnel talked them out of that and indeed, it didn't take all that long for the train to Baltimore and points south to pull in. Loutje, Wim and Frits boarded the train and this time they had a sleeper. Neat, very neat! "This is what you call a PULLMAN, I believe." "I believe, I believe, I do believe, indeed!" They all believed!

"Lalalalala. Dinner In The Diner, Nothing Could Be Finer,

Than To Have Your Ham And Eggs In Carolina!" They weren't in Carolina yet, but it was oh so romantic, just the same." The train menu was as good as any restaurant in Holland would have been, so they ate with gusto.

There was nothing mysterious going on, like on "THE ORIENT EXPRESS" and nothing glamorous attacked them either, but it was a good ride and an enjoyable diner. The dullness of the dark landscape gliding by and the clickety-clack of the wheels made them drowsy and they sacked out early.

In the morning, after "HAM AND EGGS" in Carolina, the train stopped in Goldsboro at a little bitty station. No Air Force bus was in sight yet so they lingered around and saw things they had only read about, but didn't believe existed. Water fountains and bathrooms labeled: "WHITES ONLY" and "COLORED ONLY". They gaped at the signs, not quite believing what they saw. It was true enough, though. "Welcome to the South!"

The bus showed up and 40 minutes later they pulled into the gates of STALLINGS AIR BASE, KINSTON, N.C., their home for the next 7 months.

CHAPTER V
THE T-6, THE YELLOW MONSTER

"YOU ARE NOTHING! YOU HEAR? YOU ARE NOTHING! YOU ARE PRE-FLIGHTERS AND THAT MEANS YOU ARE NOTHING! I WANNA HEAR: 'YES, SIR!' LET ME HEAR IT!"

"YES, SIR!"

"LOUDER ... LOUDER!"

"NOW DON'T YOU FOREIGN STUDENTS GIMME THAT SHIT: 'NO SPEAK', CAUSE IF YOU COULDN'T UNDERSTAND ENGLISH, YOUR COUNTRY WOULDN'T HAVE SENT YOU OVER HERE, SO WE DON'T BUY THAT SHIT. DO YOU HEAR?"

"YES, SIR!"

"I CAN'T HEAR YOU!"

Louder this time: *"YES, SIR!"*

The introduction was immediate. They barely had time to drop off their luggage before being marched off to the gym. There, a group of cadets, about the same age, stood on stage and one after the other faced the mike and spoke or screamed or hollered. This one was hollering:

"FOR THE NEXT SIX WEEKS YOU ARE RESTRICTED TO THE BASE AND YOUR BARRACKS. WHEN YOU HAVE TO GO SOMEWHERE ON THE BASE YOU DO NOT WALK, YOU TAXI. YOU RUN WITH YOUR ARMS OUT WIDE, ZIGZAGGING ALL THE TIME. YOU SALUTE EVERY UPPER CLASSMAN YOU

SEE, PLUS THE BASE OFFICERS. WHEN YOU ARE SPOKEN TO BY AN UPPER CLASSMAN YOU SNAP TO ATTENTION, SALUTE AND SAY: 'SIR, CADET ... SAY YOUR NAME ... SIR!' IS THAT UNDERSTOOD?"

"LET ME HEAR THAT 'YES, SIR' LOUDER THIS TIME."

"YES ... SIR!!"

"WHEN YOU SCREW UP YOU GET GIGGED. WHEN YOU FORGET TO TAXI, SALUTE OR SAY: 'YES SIR', YOU GET GIGGED. IF YOUR BED IS NOT MADE RIGHT, IF YOUR ROOM IS DUSTY, IF YOUR TOOTHPASTE ISN'T ROLLED RIGHT, YOU GET GIGGED. IF YOUR SHOES AREN'T SHINED, IF YOUR GIGLINE IS OFF, YOU GET GIGGED."

He showed how his fly, his belt buckle and shirt-line are all in one straight line.

"WHEN YOU HAVE 6 GIGS IN A WEEK YOU WALK A TOUR. FOR EACH ADDITIONAL GIG, YOU WALK AN ADDITIONAL TOUR, SO IF YOU HAVE 10 GIGS IN A WEEK, YOU WALK 5 TOURS. A TOUR IS A 50-MINUTE WALK AT ATTENTION WITH TIE AND WHITE GLOVES. YOU WALK 50 PACES, MAKE AN ABOUT-FACE AND WALK BACK 50 PACES. YOU DO THAT FOR 50 MINUTES. AFTER YOU HAVE WALKED ONE TOUR IN THIS HEAT AND SURVIVED IT, YOU'LL NEVER WANNA WALK ANOTHER ONE IN YOUR LIFE. IF YOU ACCUMULATE TOO MANY GIGS, YOU'LL GO BEFORE THE BOARD AND YOU MAY BE KICKED OUT OF THE CADET PROGRAM. AFTER PRE-FLIGHT, YOU'LL BE FOURTH CLASSMEN AND YOU'LL SUFFER A LITTLE LESS, PLUS YOU'LL HAVE THE RESPONSIBILITY OF MAKING THE NEW PRE-FLIGHTERS MISERABLE. THEN YOU'LL MOVE UP TO THIRD, SECOND AND FIRST CLASSMEN AND YOU'LL BE UP HERE, HELPING TO RUN THE PROGRAM. ALL THE WHILE, YOU ARE SUBJECT TO GIGS IF YOU SCREW UP. DO YOU UNDERSTAND?"

"YES, SIR!"

"LOUDER".

"YES, SIR!!"

"NOW FALL OUT, LAST ROW FIRST AND TAXI TO YOUR QUARTERS! LET'S GO!"

Back in the barracks, sitting on their bunks, Frits asks: "How often do I have to go through these chicken-shit programs?"

"Till you're a pilot."

"God forbid. I'm tired of it already!"

They received some support from the Dutch upper classmen. "Just do it and they'll leave you alone."

They're taught how to hang their clothes, one inch apart; overcoat, blues, khaki's, shirts, civvies. All buttons buttoned, even the pants hanging underneath a jacket or a shirt. All zippers zipped, of course.

Top drawer, towel straight across the bottom. Toothbrush clean, one inch from the edge. Toothpaste, rolled tight, one inch from the toothbrush. Soap, clean, and on and on and on.

"Go to the P.X. and buy black socks and halters."

"Are you kidding?"

"No, and buy collar-stays."

"What are "collar-stays?"

"Buy a nail clipper and … "

"What's a nail clipper?"

"Buy a shoeshine cloth."

"MY GOD, I THOUGHT WE CAME HERE TO FLY?"

"Shut up. Just do it and you'll get plenty of flying time."

Well it wasn't easy on his mouth, having to stay shut, but he managed. He even got a kick out of the taxi-ing and the new haircut, virtually bald, didn't look bad on him at all (he thought).

Most of the bull that was thrown at them he could shrug off or live with, but some of the fourth classmen that had just lived through that crap enjoyed taking out their frustrations on the new kids, and Frits made up his mind to kill a few of them.

Classes were fun. Military, Meteo, Engines, Instruments, Morse code, Navigation and Aero-dynamics were all right up his alley. One class startled them thoroughly. A class about V.D. Either he was very ignorant, or Holland was a very clean country. Sexually speaking, that is.

He had never heard of Venereal Diseases. It had never been discussed in the Netherlands Air Force either. In North Carolina though, they watched gory films about all sorts of diseases and shown maps, where V.D was rampant. The states were either in white, shaded or black, showing the various degrees of infection percentages. North Carolina was in *black!* Meaning that the incidents of V.D. were amongst the highest in the nation! What a bummer. The graphic pictures of the diseased soldiers nearly made them throw up. It scared them enough to consider celibacy!

There were four groups classes of about 27 each, for a total class of 110 cadets. 80 Americans, all with a minimum of 2 years of college, 25 Frenchmen, 2 British Officers, 3 Norwegians, 2 Danes, 4 Italians, 2 Belgians and 3 Dutchmen. The Northern Europeans, the Dutch, Belgians, Danes and Norwegians all spoke good English, but those Latins ... !! The French and Italians had a tough time with the English language.

The Italians were comical. Nice, but comical. Two were from "ROMA," the big city, PAOLO BRANDIZZI and GIORGIO MORELLI, one built like a boxer, the other like a beanstalk. They taught dirty words the "ROMAN" way, not the "ITALIAN" way.

GIOVANNI CAMANA was a gentle soul from BELUNO, north of Venice. He was devastated over the fact that they cut off all his black wavy hair, his pride and joy. Antonio de Blanco was from MILANO and he spoke French. Antonio and Frits could communicate in French and in turn, Frits could translate into English and vice-versa.

If any of them had a problem, they talked to Antonio and he talked to Frits in French and he, in turn, talked to the Americans and was supposed to relay the solution.

Frits picked up a lot of Italian that way and helped the boys with their English and that's what made it so comical. They had a crash course in English before leaving Italy, but no one taught them American. In their minds, "*COME VA?*" meant, "*HOW DO YOU DO?*"

The problem; no American asked "*HOW DO YOU DO?*" They asked, "*Hey, whadda ya say?*"

"*I say noseeng*" because he didn't say anything.

Or: "*Giovani, what's up?*"

"*UP? Ze sky! There!*" And he'd point to the heavens.

Or it could be: "*Giorgio, what's cooking?*"

"I no know, I no eat yet!"

Then they'd come running ... "*Meester Freets, Meester Freets, leesten to me, leesten to me! CHE STUPIDI AMERICANI!*" And they'd rattle in Italian, expressing their frustration.

One thing compensated them for their frustrations, MONEY! They had first class uniforms of the finest materials and they received more money than any other cadet: $300 a month.

Studying hard

Americans: $108, French: $90, Danes: $76 and the Hollanders: $72, lowest on the totem pole. The Italians could live like kings.

At night, locked up with other guys in his alphabetical bracket, turned out to be fun. His neighboring bunkies were "FINCH" and "FLORY", leading to a change in lyrics to that old tune: *"It's Still The Same Old Story, A Fight With Finch And Flory!, A Case Of Do Or Die!"*

Another "bunkie" across the aisle unified the whole bunch.

Merle Andrews had been an assistant choir director at the University of Illinois and he organized a choir. During six weeks of being locked up, Frits learned all sorts of old songs, most of them had never made a "Hit-Parade", so consequently, he had never heard of them.

"I've Got Six Pence", "I've Been Working On The Railroad", "Eyes Of Texas", "Down By The Old Mill Stream", "Sweet Adeline", "Wait Till The Sun Shines, Nellie", "We're Poor Little Lambs", "The Bells Are Ringing, For Me And My Girl", "And It's Beer, Beer, Beer, That Makes You Feel So Queer, In The Corps", "Off We Go, Into The Wild Blue Yonder", "Junior Birdman", "Anchors Away", "From The Halls Of Montezuma", "My Coney Island Baby", "Paper Doll" and on and on and on.

Sharp looking cadets

They'd sing by the lamplight, night after night. He loved that part. Besides, the Dutch Upper Classmen were right: "Just do all the chicken-shit things they tell you to do, don't bitch and time will fly by.

The Saturday morning inspections were the main headaches, the main "GIG" producers. First: the STANDBY inspections. In the barracks each Cadet standing by his bunk, while a white-gloved Upper Classman checked every ledge, windowsill, toothbrush and sock for reasons to "GIG" the Cadet. Next the outdoor formation inspection. It seemed that if the Upper Classmen wanted to find a speck of dust or a crease that wasn't sharp enough, they always managed to find one.

Then the parade. The presentation of the colors, the marching, the standing in the sun, all those things he despised with a passion. How many years had he been going through this crap? What had that to do with flying, shooting down MIGS?

He found a solution. He joined the "DRUM AND BUGLE CORPS." That meant practice one night, outside of the barracks. Plus, after PRE-FLIGHT, an extra night off base.

Now, as a drummer, he could walk over to the Cadet Club instead of lining up for inspection and after leisurely drinking a nickel soda, they'd stroll onto the drill field, drum under one

Flory, Frits, Flynn
Off base for the first time

arm, chair under the other and they'd set up and relax until the troops appeared in the distance and they'd start picking up the beat.

The American flag neatly hid the chairs and the drums, and for about 15 minutes they drummed, paused as the roll was called, then rolled the drums as the flag was raised and played a beat again as the troops marched off the parade field. What a job! The poor Cadets went through hours of torture, standing at attention during inspection, forming up and marching to and from the parade ground, while the drum and bugle corps worked for about 15 minutes.

All there was left to do for them, after the troops disappeared out of sight, was to pick up the chairs and drums and walk back to the Cadet Club to have a soda. Tough duty!

The bandleader was a clown and had lyrics for every beat they played. The one that gave Frits' left hand a fit was the Samba beat. It made for a "CRISP" marching beat, so in order for Frits to develop his left hand rhythm, the band leader had him sing to himself: "*Suck My Tit, Bite My Nipple Off, Suck My Tit, Bite My Nipple Off*!" It worked ... the drumbeat that is. He never did present him with a *TIT*.

Labor Day 1952 arrived and the base was closed for the long weekend. The PRE-FLIGHTERS got a break. They got "OPEN POST" from 10 am to 6 pm.

"WHOOPEE! WHERE TO?" Andrews convinced Frits to go and play golf with them. He had a car and clubs. They would teach him. *"DON'T WORRY, IF YOU CAN SING, YOU CAN PLAY GOLF!"*

The important thing was, *THEY WERE OFF THE BASE!* Frits was in for some surprises, though. The first one was when they stopped for gas, 29 cents a gallon, at some rundown, paintless shack. The shack had a cooler and Frits saw the word BEER. Root beer. He didn't know root beer from Pabst, so when he opened the bottle and took a sip, he spit the whole mouthful all over the counter and the chewing-gum rack. *"WHAT THE HELL IS THIS?"*, he was choking.

"THAT'S ROOT BEAH!"

"That's no BEER!" He gave it away.

Another surprise was the golf game. The Clubhouse was majestic, the grounds were beautiful, but the game was something else. With proper instruction he hit the ball pretty well in the direction in which he aimed it. That was no problem. Putting was not bad either. He shot 11 on a par-5 green and the guys assured him that that was not bad for the first time out.

When they lined up at the next TEE, he asked; "Now do we do this all over again?"

"Yes."

"When does the action start?"

"This is it!"

"This is it? This is all we do? I'm a soccer player. I run at full speed for two 45-minute periods. That's a sport! I'm too young for this, call me when I'm 50! I'm going for a beer!"

He walked back to the Majestic Colonial Clubhouse and there they hit him with another surprise.

"WE CAIN'T SELL NO BEAH HEAH, THIS IS A DRY STATE!"

"A what?"

"We cain't sell no beah or likker in this heah state!"

"The whole state?"

"Whea y'all from?"

"I'm from Holland and I'm going through Cadet training

at Stallings Airbase in Kinston, and I haven't had a beer for a month!"

"We-ell, you cain't get one heah either."

"*DRY STATE*?" he muttered to himself, "Give me a coke please."

A foursome at a nearby table had overheard the whole thing and waved him over to their table.

"Git yesef a gla-as en some ahce, buddy and come back."

He did as he was told and out of a brown bag on the table they poured him some Kentucky Bourbon on his AHCE cubes and told him to pour his coke over it. That was better, nearly like rum-and-coke.

His new-found friends explained the laws in the southern states, the dry states, how they could go to clubs and dances and order "SET-UPS" and "BROWN BAG" their own booze in. At the Golf Club and other private clubs, they could keep their bottles in their lockers along with their golf stuff and if they ran short they could call a cab and get some!

"A cab? A taxi? Where does he get it?"

"He buys a supply from the A.B.C. stores on Saturday, for Sunday shortages, and he makes a few bucks that way."

"A.B.C. store?"

The questions and answers kept coming about A.B.C. stores, "MOONSHINE", "WHITE LIGHTNIN'" and finally Frits asked; "Then why do you call it a dry state? You people drink more than we do in Holland with a bar on every corner."

They laughed their fool heads off. "This heah boy is pritty damn quick!" More uproarious laughter.

Then the questions turned to Holland, to flying, the Cadet program and they were having themselves a fine ol' time, half full of booze, when the busy golfers showed up, soaked with sweat and sober. Frits was "cool" and full of bourbon.

The "boys" were asked to drag up some chairs and have a "TASTE" and soon the jolliness spread in direct proportion to the Bourbon. Two shots each and it was time to go. They couldn't leave though, until they gave a BARBERSHOP rendition of "*Nothing Could Be Finah, Than To Be In Carolina In The Morning!*" They made friends for life.

All the way back they sang at the top of their lungs and they continued well into the evening, *RADEDOODAH!*

Ground school was soon being complemented with Flight training. The instructors were civilians and each instructor was assigned 5 students ... usually 3 or 4 Americans, and 1 or 2 foreigners.

Frits' instructor was "JOHN T. McGOUGH." He worked for the contract flying school, totally run by civilians, most of them former military pilots. McGough (he pronounced it "McGoo") was from Alabama, had a terrible southern drawl, had flown 34 months off a carrier in the Pacific during the war, had never made a bad landing and never got one hole in his plane. In other words, he was so perfect he was impossible!

He demanded perfection from everybody and drove everybody crazy. He had one French student who asked for another instructor after just two flights, and three Americans who asked for transfers before they had 10 hours.

Frits nearly left his group and nearly got kicked out of flight school before he had 15 hours. McGough had this terrible habit of screaming in their earphones when they did something wrong. *"I 'ONT THAINK, YO AIN'T GOT NO DAREJAM BIZNIZ IN THIS HEAH FLYIN' BIZNIZ, I 'ONT THAINK!"*

Instead of correcting or suggesting when a mistake was made, he'd start that scream again, leaving the startled student wondering; "What the hell did I do this time?"

At 3,000 feet they had to perform the "3,000 foot" check, mainly consisting of reducing the MIXTURE setting and a few other things.

On this particular day, while the race to "SOLO" was on everybody's mind, Frits was climbing out, right on course, exactly at the right speed, everything "EXACT", except he was already at 3,100 feet, not thinking about the 3,000 foot check, when that voice hit him in the ears; *"I'ONT THAINK YO AIN'T GOT NO DAREJAM BIZNIZ IN THIS HEAH FLYIN' BIZNIZ, I 'ONT THAINK!"*

"OH, SHIT! 3,000 foot check, 1, 2, 3, 4, nothing to it. DONE!" That's what he thought!

After what Frits thought was a pretty good flight, he was told to write the 3,000 foot check 5,000 times this week-end and turn them in on Monday morning. They had the long weekend off. No inspections, no marches ... JUST OFF ... FREE! From Friday at 6 PM until Sunday at 6 PM, first long weekend off.

5,000 times? What a bummer!

He started to write and after a few minutes, decided to time it. He wrote it out 10 times ... TIME ... 17 minutes ... DAMN! Calculate ... "MORE THAN 14 HOURS OF STEADY WRITING. That's crazy! I'm not doing it!"

He packed it in and went out on the town. On Monday morning he nervously faced McGough.

"Did you do your assignment?"

"Sir, I timed it. It would have taken me eight hours a day for two days in order to complete it, and it was a long week-end off and I didn't do it. So I can ask for another instructor, or you can throw me out, or whatever you wanna do, but *NO, I DIDN'T DO IT!"*

McGough was flabbergasted and got very pale. His lips barely moved when he said; "Let's go see the Flight Director!" The Flight Director and Contractor of the Flight school left both of them standing as he listened. McGough didn't say much; "I gave this boy an assignment and he refused to do it."

Frits told his side of the story, showed him how he had timed it Friday night and was excused from the room.

After McGough left, Frits was called back in. "Fri-its, I ain't condonin' disobeyin' orders, you unnersta-and? But I do think that was severe punishment for what you did and that was uncalled for. You have a fine record so far and if you want to fly with another instructor, I'll transfer you."

"I wanna stay with Mr. McGough, but I don't wanna write for 16 hours, Sir".

"O.K., do you wanna tell him about your decision, or shall I?"

"I'll tell him, Sir."

"Alright boy, keep'em flying!"

He walked back to the Flight Room and sat down in his usual seat with McGough and his students. McGough lifted an eyebrow in question and Frits nodded and not a word was said. The whole affair was never mentioned again by any of them.

"GOING SOLO" was the main topic and the main goal of every Cadet. NOT solo-ing meant "WASH-OUT" and solo-ing early meant real "AIRMANSHIP". All the training revolved around getting them ready to "SOLO." Nearly every night, in the Mess Hall, the bell would be rung and after everyone hushed, the bell-ringing Cadet would shout "Hear ye, hear ye, I'm Cadet so and so, and at ... blank ... hours, I solo-ed, making 3 landings, no "GO AROUNDS!"

Then, of course, everyone applauded and shared the Cadet's moment of glory. Sometimes there were 3 or 4 such announce-

Winter flight outfit

ments a night.

The average time to SOLO was around 24 hours, so anyone who SOLO-ED at 20 hours or less was a real ace, and the next important thing was NO GO-AROUNDS. When on the final approach to the landing, not everything CLICKED: the speed, the height, the angle to the runway, the pilot would be well advised to GO-AROUND rather than risk a bad landing. Of course ACE PILOTS made no mistakes, so NO GO-AROUNDS.

Landings were made at three airports. Kinston, of course, but most of the practice landings were made at AUXILIARY airports, such as Goldsboro and Greensboro. Each had a fair-sized runway and very little air traffic. The instructors would leave in an airplane in the morning, along with one of their students, and the rest would be put on a bus and driven to the auxiliary field. There they would hang around and watch student #1 do his "TOUCH AND GOES" until it was time for student #2 and so on. Most students got to fly about once, every other day. That wasn't too bad.

During that period there was a break caused by a hurricane threat, and all instructors and SOLO students climbed into a plane and left for Virginia, well away from the path of the storm. The base was deserted for four days. The only ones around were the lower classmen and the hurricane never hit. It was like having time off on base. PEACEFUL, VERY PEACEFUL!

The T-6, or TEXAN, was a big low-wing bird, with a big radial engine and an adjustable prop. Much heavier and faster than the little TIGER-MOTH. It had dozens of instruments instead of just 3 or 4 and had switches, like a commercial airliner. In class they learned all the principals and theories behind the switches and instruments, in the air, the practical side of it.

Keeping all the needles on all the gauges where they were supposed to be was the trick. Having the BALL centered in the "NEEDLE AND BALL" was crucial because that meant they were flying coordinated. That was more easily said than done. On take-off, the torque of the engine under full power wanted to pull the aircraft to the left, so hard right rudder was required to keep it going straight. In a power-off glide it was just the opposite. Other than that, things were fairly simple.

To maintain level flight, all they had to do was keep the wings level, the ball centered, the altimeter at the same altitude, the compass heading the same, the vertical speed indicator at zero, the RPM correctly set, the trim-tabs set just right, the mixture in the right position, the oil temp and pressure correct, the air-pressure reading right, the cylinder-head temp right, the suction gauge reading properly, the correct heading on the radio-compass, the VHF tuned to the right station, the wind correction calculated properly, the variation figured in, the deviation as well, and otherwise they had little to do but check the ground versus their course and the sky for other aircraft. Of course, they had to also know where they were going.

When they had to change course or altitude, roll the plane or change the pitch in order to glide or climb, other factors start playing a part. The compass started to roll ahead or lag behind, a bank caused loss of lift and required additional throttle, on and on and on.

The whole idea was that all that became second nature, so that these youngsters, a year later, could climb on the tail of a MIG 15 over Korea and concentrate on the job at hand: *BLAST THE MIG OUT OF THE SKY*!

For some it proved too much of a burden and they "washed-out". Others just kept flying along.

Frits found the T-6 a "bitch" to fly compared to the TIGER-MOTH, but Mr. McGough kept flying him, so he must have been doing alright.

On a sunny morning in November, the bus to Greenville was full of enthusiastic, young aspirant pilots. One of the British Officers was telling TEXAS jokes, with what he thought was a Texan drawl, and the combination of his COCKNEY English and TEXAN drawl made everything he said hilarious.

That was good therapy because competition was getting fierce. That FIRST SOLO was getting close for most of them, because most of them were approaching 20 hours or had passed that already. It was starting to work on their nerves. No solo, no flying career. If they didn't solo by the time they had 24 hours, a special TEST-RIDE would ensue that might allow them a little more time, or boot them right out of the program.

If that was not building enough pressure already, the race to be the first Cadet of Class 53F to stand up in the Mess and ring that bell was probably the real ulcer-causer.

The applause of 400 of their peers is a reward beyond anyone's imagination, and it's worth ... all the work, aggravation and anticipation. Everyone wanted to be the NUMBER ONE, the big cheese, Numero UNO! It was a safe bet that every Cadet was competitive as hell, because that was what got them there in the first place.

So, on the 45 minute bus ride in spite of the jokes, they were reviewing "down wind checks," "landing procedures," "radio failure", and dozens of other emergency procedures that they might have to fall back on, just in case. Knowing them inside out, forward and backward, was essential. If they screwed up, up there, they might kill themselves. If they didn't know them on the ground, the instructor might not let them fly that day.

In that frame of mind they arrived at Greenville Airport and started hanging around, criticizing the students in the pattern, doing their "touch and go's".

The sideline commentary was educational and funny, like *"GIVE IT POWER, YOU STUPID BASTARD, YOU'LL END UP IN THE WIRES!"* Or *"Pull those flaps up, dummy!", "Good God, that instructor will eat his ass!" "Look at that idiot up there, he's gonna fly the whole pattern with his gear down! And he's wondering why the plane won't climb!" "Oh God, look at that! He's gonna stall out on final!" "Dumb ass, put that nose down!" "Hey, good show! Who's in 483?" "He's alright!"*

Obviously, all the experts were on the sidelines and all the "failures" were up in the air. That scene was repeated on least 50 training bases throughout the U.S. Young, eager students thoroughly enjoying themselves.

When McGough came down with his student for the last time, they taxied to the ramp and after de-planing, McGough de-briefed the man for about 15 minutes. He went for a leak and some coffee, it was time for Frits' turn.

"MR. FORRER, I AIN'T SAYIN' NUTTIN IN THE AI-AIR. YOU JES SHOW ME THREE PERFECT LANDINGS. FIRST ONE, YOU TURN OFF, SECOND, YOU TOUCH AND GO AND THEN TURN OFF AGIN. I AIN'T SAYIN' A WORD. YOU KIN SCREW UP AND YOU KIN DO RAIGHT, THAT'S UP TO YOU."

Well, that was clear enough!

After inspecting the plane, Frits climbed aboard, went through his checklist, called "*CLEAR!*", and started the engine. After a radio check on different channels he settled on channel

2 for Greenville! He taxied out, did his engine run-up before lining up on the runway and, when cleared, lined up on the center-line and with a lot of pressure on his right rudder, he pushed the throttle forward.

A quick glance around the cockpit, the canopy open, all clear out front, glance back at the gauges, full power, more right rudder pressure, tail wheel up, keep it going down that centerline, gauges O.K., slight back pressure on the stick and ... she's flying. Gear up! Set up climbing speed, milk up the flaps, more back pressure, look over the shoulder, straight out, good going, speed at ninety, reduce power and pitch a little, maintain ninety, power at 2300 RPM, clear on the right? 400 ft, right turn, maintaining 90, roll out 90 degrees, clear to the right, roll right again, watch that speed, pick a point over the wing, level off, check speed, reduce RPM, fine, 800 ft. all clear on down wind, keep her at 800 ft. exactly, speed is picking up, keep that nose down, power back to 2100, check distance from runway, he kept talking to himself. Check for wind drift, good show, parallel to the runway, maintaining 800 ft, across from the end of the runway; power to 2000, carburetor heat OUT, mixture FULL RICH, pitch control, FULL FORWARD, speed 110, 10 degrees of flaps, trim nose up, speed to 95, right turn 90 degrees, 20 degrees flaps, 95 steady, right again, roll out, straight in front of the runway, looking good, full flaps, keep that nose down, check that speed, down to 80, cut power, looking good, over the overrun, pull back slowly, slowly, round her out, slowly, keep her straight with the rudders, pull back, pull back, pull back, all the way into the belly ... TOUCHDOWN on all three's.

Carburetor heat in, flaps up, taxi to the left at the next turn-off, "O.K. so far!" Two more times, same results, except after the second landing he gave it the gun, and took her around without leaving the runway.

After the third landing; *"TA-AKE ME BA-ACK TO THE RAMP"*, the first words spoken since they started.

While parking the voice came on again, *"KEEP HER A-RUNNEN'."* McGough climbed out, and while standing on the wings, reached into the cockpit and fastened his shoulder straps over the empty seat in the rear, tightened them and closed the canopy.

Leaning into the cockpit in order to avoid the idling propellers blast, he screamed over the sound of the engine;

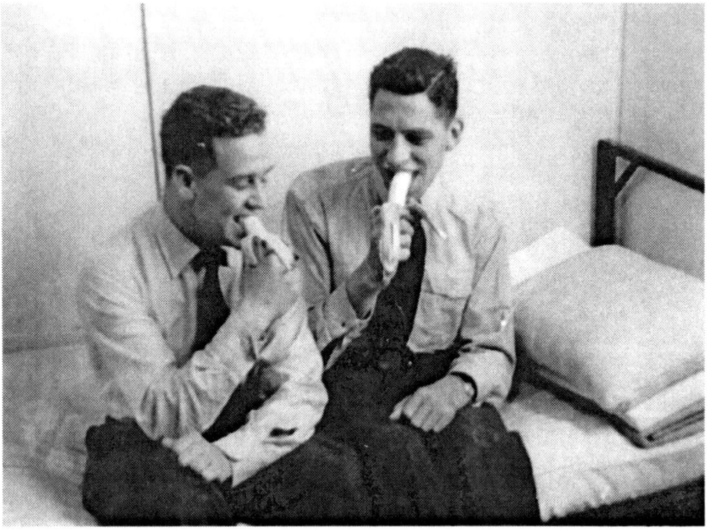

Frits and Shorty go bananas

"NOW, YOU KIN TURN MAH FAICE AS RED AS DAT DERE FAHRTRUCK BAH MAKIN' SOME BAD LANDINS, BUT AH EXPECT YOU TO GIVE ME THREE GOOD'UNS. THREE TOUCH AND GO'S AND COME BACK!"

With that, he walked down the wing and Frits was alone. The biggest difference was the weight. No one in the back seat, compared to the TIGER-MOTH, nobody in the FRONT seat. The tail came up sooner, the plane left the ground faster and he had to watch his speed when coming in for landing. He throttled back a little sooner on down wind and lost an additional 50 feet before turning base leg. While in the final turn he dropped FULL flaps, rather than waiting till he completed the turn. It worked nicely, he touched down at the second marker, just like he had the three previous times and after a quick "CARB-HEAT" IN, FLAPS UP, he gave it full throttle and repeated the whole process two more times.

When he "ramped-out" this time and turned off the engine, McGough came running up the wing, slapped him on the back and grinning an unusual grin from ear-to-ear, *"AH KNEW YOU COULD DO IT, YOU MISSABLE SON-OF-A-BITCH. YOU DONE JUST FAHN, CONGRATULATIONS, FIRST IN YO CLASS!"*

T-6 Pilot?

That night, at chow, was one of the proudest moments of his life!

DING, DING, DING, and again, *DING, DING, DING! "HEAR YE, HEAR YE! TODAY, I, CADET FORRER, FIRST ONE OF THE CLASS OF 53F, SOLO-ED, THREE TOUCH-AND-GO'S, NO GO-AROUNDS, AT 18 HOURS AND 20 MINUTES!"*

The applause, the back slapping, the handshakes, the complimentary remarks, and congratulations and all the questions had him flying higher than a kite! All the aggravation, all the chicken-shit, all the gigs, the tours, the complete lack of girls, the lack of beer and dancing, it was all worth it! Today's announcement in front of 400+ fellow flyers offset any disappointment or misery he had endured. IT WAS GRAND!

The Dutch contingent, with whom he was sitting, acted in the most enthusiastic fashion. Even the PRE-FLIGHTERS, who weren't supposed to talk at the dinner table let alone jump up, came over to backslap and congratulate him. It was a grand, great experience!

After chow, when they stopped by their mailboxes on the way to the Cadet Club to celebrate the occasion, he had another surprise, LETTERS AND CARDS!

IT WAS NOVEMBER 14, 1952, HIS 21st BIRTHDAY!

CHAPTER VI
NOTHIN' COULD BE FINER......

The next shipment of cadets brought in FIVE DUTCHMEN, as different in size and personality as anyone could possibly have assembled.

The tallest one was a born clown and at 6 feet 4 and 230 pounds, he was in great shape. In Dutch he was called; "DE LANGE," (tall one), but in America he was immediately named "SHORTY."

He was also the oldest and already a Corporal, having finished training for "wireless-operator".

He could curse somebody in *'DADADIT—DITDITDA-DADADIT"* so rapidly, even the instructors couldn't follow him.

"Jan Elberse," tall and good looking, thought he was "BING CROSBY" or better and wished he were in Hollywood.

"Ge Eijkelenkamp," nick-named "Ike" was also 6 foot plus and tough talking, spoke with a thick East Holland's accent. He had been a "GROLSCH-BEER" truck driver in his father's business in the eastern town of Deventer.

"Kees Bos" could have been Frits' twin. The son of a Minister, he had the same build, hair and face and he was full of the devil.

"Pauw" was short, stubby and studious, more quiet than the others, but he could "BURP" so loud, he'd stir up the whole movie theater.

The new gang got to go out together for the first time at an unusual occasion.

All the FOREIGN STUDENTS were invited to march in the HOMECOMING PARADE at EAST CAROLINA COLLEGE, (Now East Carolina University) in Greenville N.C.

That seemed like a blast!

Unfortunately, they were escorted by one of the few Regular Air Force Officers on the base.

He had them "line-up" and wait, rather than "mingle with the students" and wait.

About 80 Frenchmen in their black "dress uniforms," 10 Dutchmen, 4 Italians, 4 Belgians, 6 Danes and 7 Norsemen made for a colorful group of young men.

The college girls seemed to think so too, yet the cadets had to stay in line and all they could do was smile. Damnit.

When the parade started up, the French contingent sang one dirty marching song after the other, while the American Lieutenant smiled broadly at the cheering crowd.

He had no idea, what was going on and it would have been embarrassing, if the people on campus had understood what they were singing

It was like singing on crowd-lined streets; *"My Shoes Are Loose, My Pants Are Tight, My Balls Are Swinging Form Left To Right.*

Sound Off, ,,,Sound Off!!!!!

Or....;*"I'm Looking Under A Skirt And Wonder, Why I Never Looked Before!"*

It was hilarious and the Dutch, Belgians and Italians chimed in with them.

Finally, after the parade, they were turned loose. They were assigned Hosts and Hostesses, were escorted to the cafeteria and then to the "HOME-COMING-GAME".

In Holland, "SOCCER" is known as "VOETBAL", "FOOT-BALL' in England, so "FOOTBALL" should have been entertaining to them!

WRONG! Frits' first question was; "When are they gonna start doing something?"

From his viewpoint; he saw 22 athletes in the most "unsportsman like" outfits walk out onto the field, stand around in a circle and then line up like bulls, ready to shove the horns to one another. Then everyone fell down and the crowd roared!

They got up, stood in two circles again, lined up and fell down again.

"When does the action start?"

"We would have been up and down that field twice already with a soccer ball."

"What happens next?"

His host and hostess, tried very hard to make him understand the game, but after 20 minutes he excused himself. There had to be a little bar in town someplace!

Wrong again! Not in Greenville!, Not in 1952!

He strolled the town, wondering how the other Foreign Students were enjoying the game.

In the Hotel lounge, over a cold Coke, he got to know a little about the town and where there was a band at night. He was advised to get a "Brown-Bag" from the A.B.C. store before it closed and that sounded like a good idea.

Not having a driver license and not having his pass-port with him, he couldn't prove he was 21 and they refused to sell him anything. The store manager insisted that all military bases issued I.D.'s for their troops and didn't believe that STALLINGS hadn't issued any. When he told them that in his country you could walk into a bar, unescorted, at 15 and order a beer, he could see in their faces that they pegged him as the biggest liar in town.

DAMN, DAMN, DAMN. Back to the campus.

At least, by then, they were partying all over the campus and he did latch on to a FRAT-HOUSE that was celebrating the FOOTBALL VICTORY with a keg of beer. The students had no problems accepting a Dutch Cadet in their midst and Frits had no hang-ups about singing songs with them with a beer in his hand.

They were a fun bunch. A little too bent on getting stone drunk, but a fun bunch just the same.

It took 40 minutes to kill the keg and the party fell apart, with the participants drifting off to other parties.

Frits ran into the other cadets and was informed that they were being treated to dinner and a dance afterwards.

The Air Force bus would leave at 11 PM, so that left them

a lot of time to get acquainted.

There were a lot of pretty girls on campus and every guy did his damnedest to get paired off with one. That wasn't particularly easy. Most girls already had a date or even a "steady", but when the band struck up; *"In The Mood"*, he found a great dance partner.

"RITA" (he thought she said "REEDA"), was from Washington, N.C., not far from Greenville, toward the coast.

She talked funny, but so did most of the others, so he figured he'd get used to that.

They moved from the jitterbug to a slow dance, nice and close, that really started to warm his cockles, but came to a sudden halt.

Someone tapped on his shoulder and RITA danced away with him.

"WHAT IS THIS?" He was standing in the middle of the dance floor all alone amidst all the other couples, while his girl danced away with some yo-yo. He was flabbergasted.

He walked back to the punch bowl, madder than hell.

He told IKE (Eijkelenkamp) of his experience and Ike asked; "Why didn't you slug the son-of-a-bitch?"

"She just danced away with him!" He sounded like he still couldn't believe it.

Giorgio Morelli joined them and his; *"CHE STUPIDI AMERICANI!"* was a good indication, that he had a similar experience.

"CHE FIGLIO DE LA MIGNOTA!", he kept cursing.

Ike wanted to know what he was saying.

"He's saying: "That son-of-a-bitch" in Roman!, They must have done the same thing to him."

"Roman? You mean Italian? Right?"

"No, in the rest of Italy they say; *"CHE FIGLIO DE LA PUTANA!"*, but he curses the Roman way!"

"Is this what these clowns teach you?" Ike wants to know, "How to curse in Italian?"

"Yes. And how to say *"I LOVE YOU!"* and *"YOU HAVE NICE LEGS!"*

Ike laughed uproariously. "I need to learn that stuff too!"

They were getting over their grief and started looking for other girls to dance with.

"REEDA" reappeared, acting like nothing happened, so after some punch, they took to the dance floor again.

Shorty acting up

Nobody cut in during their jiving, but the moment they lapsed into a slow dance, some yo-yo cut in again.

Rita seemed to think that that was normal, but Frits was fuming.

"Why can't those bastards go out on their own and ask a

girl to dance?" he asked Rita over a punch.

"I don't know, but why don't you cut in too?"

"1 don't like that system and I wouldn't lower myself!"

After getting Rita's phone number at her dorm, it was time to catch the bus.

It was the end of his first day ever at an American College.

IT HAD BEEN AN EDUCATION!

The following week-end changed their life styles.

Someone had informed someone else that the City Engineer in Goldsboro was a native of Holland and somehow, they received an invitation to meet the Engineer and the Mayor at City Hall on Saturday afternoon.

Somehow, everyone got there. Some bummed rides with classmates who had cars, some took a bus, but KEES and FRITS, dressed like twins went hitchhiking on "WOODEN SHOES".

They figured that the Wooden Shoes would be a car-stopper and it kinda worked out that way.

At City Hall, they met the Dutch City Engineer, Mr. BEEM, the Mayor, and the City photographer took official pictures and promised to send them prints.

After a tour of City Hall, the water tower and the sewage treatment plant, they were invited out for dinner at the ELKS CLUB.

The dinner was excellent, even though most Dutchmen wouldn't touch their "Corn-on-the-cob".

The same "Brown-Bag" procedure was followed at the Elks Club as had been the case in the Country Club.

Members had their own lockers in which they stashed their bottles and the club provided ice and set-ups.

Most of the guys would have preferred a beer over a "bourbon and coke", but you can't look a gift bottle in the mouth, since everyone seemed willing to share their bottles.

The "Hillbilly band," correction: the "COUNTRY BAND" played acceptable music, but the "COUNTRY" type of dancing was not necessarily Frits' favorite.

They danced their shoes off just the same, particularly with one good looking "Doris Day" lookalike. She not only looked good, she was a great lady and a terrific dancer.

She was also married and was accompanied by her parents and her brother.

Of all the breaks, she was married to an artillery Officer in Korea, but was having a good time, being out with her family. She introduced the guys to her parents, the "Robinsons", the owners of "ROBINSONS" drugstore in town, their hangout to-be for a long time.

During a break by the band, the Dutch boys were introduced and asked to sing a song for the gathered ELKS.

Well,,,,,, they decided to pull one on their audience and chose an old Dutch song ;*"T Zonnetje Gaat Van Ons Scheiden!"*

Every child in Holland learns this from Kindergarten on. It means; *"The Sun Is Leaving Us At Sundown"*.

What the clowns actually sang to it was a different story.

They sang: *"Moet Er Nog Sla Olie Wezen? Moet Er Nog Sla Olie Zijn?"*

"Is Salad Oil Needed? Must There Be Salad Oil?"

Over and over and over, throughout the whole tune.

Of course, the Non-Dutch-Speaking public had no idea, what was coming off, but the guys themselves were having a ball.

To make it even more interesting: "SHORTY" accompanied the "SINGERS" on his imaginary "BASS-FIDDLE".

Just to watch that tall man moving his left fingers up and down as if he were playing a real "BASS" with his right hand plucking at the so-called "STRINGS", while his big mouth, big red tongue flapping about, was *"THUMPING"* out the notes; *"THUMP, THUMP, THUMP, THUMP, THUMP,* was one of the most hilarious sights they had ever seen.

The boys cleaned it all up somewhat by giving a fairly straight rendition of *"The Windmill's Turning*!".

Two choruses in Dutch, one in English.

The applause was just great and just about every table in the room ended up having one of the "FLY BOYS" over for a "TAISTE" and some chatter.

Many of the men in the room had served during the BIG WAR and had been in Europe during that time.

That spawned many a conversation.

All in all, it was a great evening and although some of the guys felt no pain at all, most of them felt just "happily mellow!"

The Robinsons led everyone over to their house for drinks and songs and on their back porch, accompanied by two guitars,

(Mr. Robinson played a mean guitar himself) they sang until the wee hours of the morning.

They ended up sleeping all over the place. In chairs or beds, on couches or the floor, it didn't matter much until it was breakfast time and they were expected to take their hangovers to church.

The most terrible thing to do to a good Catholic Boy is to take him to a Baptist Church and expose his headache to a Sermon and some Psalms.

After lunch, they made it back to the base, well in time for the 6 o'clock curfew and a few more hours of study.

One Sunday, some Foreign Students were invited to the Governor's Mansion for dinner.

First, lunch at St. Mary's College in Raleigh. St. Mary's was a beautiful all-girls school, that proved to be great on one hand and frustrating on the other.

Great, because lunch and singing around the piano with all those girls was a beautiful experience. The girls were overly friendly and charming and some of them were down-right gorgeous. But it was very frustrating, because there was no time to get to know any one of them personally or even set up a date. That was tough, because the guys would gladly have adopted a few of them. Damn-it! That was too bad!

After a tour of the Governor's Mansion and the Capitol, dinner with Governor SCOTT turned out to be a pleasant affair.

A few traditional French songs added to the relaxed atmosphere and the evening was topped off by everyone receiving an "HONORARY CITIZENSHIP OF THIS HERE TAR-HEEL STATE!"

On another occasion in Goldsboro, they were invited to a Saturday afternoon BAR-B-QUE. Behind a great big white frame house, with at least 50 pickup trucks parked around it, the cookout was taking place and it was a howling success in more ways than one. The "RI-IBS" were outstanding, the cole-slaw just "GRE-AT" and the country songs "A-HOWLING".

"Your Cheatin' Heart" was sung at least 10 times and *"Why Can't I Melt Your Cold, Cold Heart"* probably just as many.

"There Sta-ands The Gla-as, Fill It Up To The Bri-im, Till My

Troubles Are Di-im" was a new one on them, but they learned fast.

Somewhere during the festivities, someone produced a stainless steel bucket with "WHITE LIGHTNING".

After pouring off half of it in another bucket and filling it back up with water, it was ready for serving.

Their first taste of homemade liquor.

All of the Dutch boys had had some experience drinking Dutch Gin, called "JENEVER", back home. They were used to drinking it straight, no ice, no mixers and no chasers, but they weren't ready for this.

Already diluted 50%, the "WHITE LIGHTNING" still burned their guts out.

Of course, they had to show off, encouraged by enthusiastic shouts like;" Y'ALL CAIN DO IT!" and "CHUGALUG" and so on. Besides that, with a cold beer chaser, they could handle it.

Well,,,,,, most of them.

Some of the "NON-SINGERS," like "KLEINE KEESJE," Keesie Bos, had much more time for drinking than the "SINGERS".

The SINGERS held their own, but let's face it, if your mouth is busy imitating ROSIE CLOONEY's *"Come-On-A-My-House-A-My-A House,I'm Gonna Give You Everything!"* or *"We Ain't Got A Barrel Of Money, Maybe We're Ragged And Funny!"*, you don't really have much time to drink too much at the same time.

So by the time the host was loading up everybody to take them over to the Robinson's, Keesje was loaded to the gills and the other guys had to kinda carry him.

While doing a very, very loud EDDIE FISHER imitation, they drove off, heaped up in a brand spanking new BUICK.

Within five minutes, Kees got sick and while the car was driving along, the guys barely got the rear window down and Kees' head out of the window before he threw his guts out all along the side of the car.

It was only 5:30 in the afternoon.

Hosing down the car at the Robinsons, Frits nearly threw up from the acrid smell, but thank God, a cold beer settled his stomach just in time.

They did their damnedest trying to convince Kees to sleep it off somewhere, but the little stubborn drunk insisted on coming along.

When everyone got out of the Robinson's black HUDSON,

the biggest car Frits had ever seen, in front of the ELKS club, Kees was sound asleep, so they just left him there, hoping he wouldn't get sick again and mess up the HUDSON.

It seemed that those Dutch boys, average age, 2O, could handle their beer alright, but WHISKEY knocked them flat on their butts, if they stayed with that stuff all night.

Dancing helped a lot. First of all, they had to put their drinks down when stepping on the dance floor.

But besides that, the alcohol came right out of their pores along with their perspiration and there was plenty of that.

Frits danced just about every dance, with young and old alike. When they did a square dance, something he didn't know yet, he was right in there learning the do-si-do and twirlin'.

There were plenty of ladies present, stuck with husbands who wouldn't dance, so our boys were very popular dance partners. When a fast polka or Vienna Waltz left the floor nearly deserted, they really had themselves room for a workout and he was having a ball.

The only problem was: they didn't get laid.

It was most difficult somehow to meet and date an available girl and end up in bed or the back seat with her.

Not having a car and not much money limited their chances severely and within the crowd they associated with, there were just no eligible bed-partners.

By December, Frits had not been in bed with anybody since he left Holland. That could get frustrating. It could even cause pimples. It was a tragic situation.

Of course, some of the stories told around the barracks would have them believe that some guys were always "lucky," but generally, that was more the exception than the rule.

Frits had heated discussions with his roommate "ED BROWN", the grandson of a "preacherman" and some of his buddy-cadets.

It was very obvious, that there were tremendous differences in their upbringing and philosophies.

The *"AMERICAN PURITANICAL* and *HYPOCRITICAL"* attitude of those guys would get his gall.

They might have been lying, but by having cars and money, they might have been in a much better position to "SCORE", but that was not the point.

The problem was their attitude. The "*DAMNED IF YOU DO, DAMNED IF YOU* DON'T!' attitude got him really tee-ed off.

If a girl did "*PUT-OUT*!", they enjoyed her, but labeled her as "*NO GOOD*" and wouldn't go out with her anymore.

If she did *NOT "PUT-OUT*", they wouldn't go out with here anymore either, because she was a "*WASTE!*".

"But how about the girl? How about her feelings? Where does that leave her? That just isn't fair!"

Frits would get so stinking mad. "Doesn't she have any feelings or rights?"

"Do you just discard her, like you discard a rubber?"

His continental mind worked differently. Especially when the attitude was; "I don't give a damn if she comes or not, as long as I knock my nuts off!"

Or: "Shit, if she gets pregnant, that's her stupid problem. I don't have to worry about that!"

That would blow his mind. The worst part was their double standards:

"My girl at home?" "Oh no! Never!" "She gonna stay a virgin until we get married!"

He'd fume: "In other words: It's perfectly alright for you guys to screw around all you can, but the one you marry has to stay celibate?"

"Yep, That's the way it is, man!"

He could just puke.

One crazy week-end, a bus would leave for Greensboro on Friday night and return on Sunday afternoon. Sleeping could be secured at the Armory or at the YMCA, if so desired.

Someone said that Greensboro was a swinging college town, with a lot of action, much better than Raleigh. So, with a toothbrush, razor and a few pieces of clothing, they were going to check it out. They would really do this town!

Four Dutchmen, surrounded by a wide array of foreign students, sang, laughed and joked their way to Greensboro on the Air Force bus.

Downtown, where the bus dropped them off, they made a real pleasant discovery: "BARS!" REAL LIVE BARS!

Why??? How???, They really didn't care. They were there and that's what counted.

Later on they learned that these bars didn't serve hard

liquor, only beer or wine, but that was O.K., that's all they wanted anyway.

"BOTTLED" beer was fine and truthfully, cold American beer could easily help them forget what Dutch beer tasted like.

In the first bar they hit, they caught the attention of two "older" girls, (at least 25) at the end of the bar, around the bend. The boys were loud and rambunctious enough to have caught anybody's attention, so catching the attention of the two ladies at the end of the bar was not particularly difficult.

At the other end of the bar, couples were seated around tables or wrestled on the little 6 by 6 dance floor to the blaring Country Music from the jukebox.

Frits asked the chubby short one to dance and learned that they were local girls, married and lonesome. Their husbands being away in the service, it did get lonesome around the house.

"And y'all?"

"Well..... we'all are cadets stationed in Kinston!"

"But what's that funny language y'all are speaking?"

"Dutch! We're from Holland."

"Holland? With all them tulips and windmills?" "And y'all stick y'alls fingers in the dikes?"

He couldn't resist that one. "The last time I tried to stick a finger in some "DIKE", she slapped my face!"

"BETTY" didn't get that.

"Tell me, do y'all really wear them wooden shoes?"

"We sure do and as a matter of fact, we've got them on base and next time we'll wear'em, O.K.?"

"Tha-at's so da-arling, I wush you would!"

Back at the bar he met her friend; "JULIE, "and after he sat down with them, some of his friends drifted over too and were introduced all around.

They danced some, joked some, laughed a lot and the beer flowed as easily as the conversation.

Once the girls established that the boys weren't tied down to any particular time and place, they suggested that they continue the party at Julie's house, because she had to relieve her baby sitter.

They made a few phone calls before leaving the bar and lined up a couple of other girls to join them at the party. The promises of girls, booze and food had everybody pile in Betty's car and off they went "A-PARTY-ING."

Two more girls joined them on the screened-in back porch, where the hamburgers were being grilled and the dancing took place. The drinking was done in the kitchen, the living room and the porch and the singing was done all over the house!

Betty had unmistakably claimed Frits for herself and that was alright by him. She had 4 or 5 years on him, had been married 3 years and was married to a career Military man, whom she hadn't seen for 8 months.

She was very, very lonely, very, very frustrated and very very hot.

Before 1 am, while the party was in full swing, she hauled Frits up the stairs, into a bedroom. Actually, she didn't have to "haul" very much,,, lead him by the pinky was more like it!

It didn't take much persuasion.

She didn't ask for or offer any kind of protection, so Frits didn't bother to worry about it either.

After all, married women should be able to take care of themselves.

Whatever she couldn't do to herself, he did to her and 6 months of pent-up frustration on his part, was just what the doctor ordered for her. She was something.... Insatiable!

Back down for more hamburgers and beer.... back upstairs for some more nookie. A guy could die this way.

Somewhere along the line HE DID!!!!

When he woke up, he barely knew where he was and his bladder was about to burst.

He had to turn 4 doorknobs before he located a bathroom, just in time to keep from exploding.

A cold beer got his brains back from "TILT" to "NORMAL" and he was brought up to date.

Julie's kids had gone off with her mother for the day, four guys were still under the roof, but two girlfriends had departed, but were expected to return for more later.

"Who wants bacon and eggs?"

"What time is it?"

"Good God. How's the beer holding out?"

Betty appeared and looked like she's been in bed with a wild man for a week.

Julie offered help; "Here honey, have some aspirin and milk. It's better than beer. Go fix your face a little and then come back!" Julie made sense.

First some food, then a joke and a lot of laughter, then a beer and the party was on again. So much for celibacy!

On the bus the next morning, he was admiring his new ring. Gold, with two diamonds and some blue stone in the center, set in white gold.

He remembers asking;" Are you sure this is a man's ring?" He wondered what he did to earn it?

At the base, life got better too. As they moved up to 3rd and 2nd classmen, they were subjected to less chickenshit, could attend movies at the base theater and sip 5 cent sodas in the Cadet Club.

The harmonizing in the barracks had gone down the tube, because nobody hung around the barracks anymore at night.

The Dutch contingent was asked to provide a half hour of song and stories for a local radio program, but they argued so bitterly amongst themselves it never came off.

Their own "Bing Crosby" insisted on singing American songs and not allowing anyone else to take charge, so he blew the whole deal.

Frits joined the French Choir and performed with them. "French is a more beautiful language anyway!" was his defense.

He bought a used 35 mm camera and from thereon in shot pictures of everything that was "different."

One of the greatest contrasts in America, to their minds, were the poverty areas.

In Holland, everyone lived in nice clean homes. Owned or rented, it made no difference. The windows were washed each Friday, the sidewalks scrubbed and the whole house repainted if only ONE blister showed.

Hardly anybody owned a car. In his immediate family, uncles and aunts or his mother or brother, NOBODY owned a car, but... ... .. *THEY KEPT THEIR HOUSE CLEAN AS A PIN.* Nobody had T.V. either!

In Kinston, strolling down a dirt street, he took pictures of dilapidated shacks without a stitch of paint, but with tall T.V. antennas and shiny new cars in the yard. The scenes would have made for a good documentary.

The water towers, so different from Dutch ones, the different planes that visited Stallings airbase, it all captured his interest and ended up on his film. A wealth of news and information to be brought back home to Holland.

To give his family a better understanding about life as a cadet, he bought an ordinary adding machine roll and started to record his daily routine.

All day long, he'd find a minute or two to write down what was happening right then and there!

From reveille to breakfast, from classes to flight line duty. Details about what they ate, studied, talked about and how the weather was holding up. How the flight had gone and the stupid mistakes he'd made, how they dressed and what movie they'd seen. He recorded the jokes he'd heard and translated them for the folks back home. In other words: an hour by hour account about their lives as cadets.

By the time he sent it, it was 49 feet long. At home they loved it.

Home was having problems though. Dikes broke in the southern province of ZEELAND and thousands of people drowned. Holland was dealing with its greatest natural tragedy in it's history.

CHAPTER VII
THAT EVERLASTING YELLOW MONSTER.

The T-6 kept chewing up his victims.

Charlie Pauw, the Belcher, washed out, two Italians were gone, just to name a few.

The curriculum didn't seem to get much harder, but some guys couldn't hack it as well as others.

None of the Dutchmen had any problems with their Academics or the Military part, but the flying got them.

ANTONIO went back to MILANO, GIOVANNI went back to the beautiful Alps and his hometown of BELLUNO, near CORTINA D'AMPEZZO. Too bad, they were great guys.

The class kept shrinking. They were heavy into navigation and cross-countries and funny things started happening.

Flying "cross-country" generally meant flying a triangle on the ground and making it back home. For instance: North-North-West to La Grange, 70 miles. Then due East to Washington N.C., and South West to Kinston. Not particularly complicated if they stayed on track.

All they had to do was fly straight and level, figure their course, speed and distances on the ground, plot it on their chart and figure in a few other things.

Their "variation", (because the "magnetic" North Pole does not happen to sit smack under the "geographic" North pole) and they had to consider that.

The "deviation": The compass itself might be off a few degrees left or right.

The "wind": That was the big factor.

If the wind was behind them, they'd go faster, obviously, if the wind was dead ahead, they'd go slower.

If the wind was off the right or left wing? That's where the humdinger came in.

From "METEO" they found out the wind velocity and direction at the desired altitude. METEO gave the wind in "KNOTS", so the Cadets would screw up some more? The airspeed-indicator read in M.P.H., the charts were in STATUTE miles, so they had to convert KNOTS to M.P.H.

Once they had the wind calculated, they could figure how many degrees they had to head into the wind in order to stay on course.

At their first checkpoint, say a railroad crossing, they checked their time and position and often found that they were to the right or left of course or ahead or behind schedule. That would indicate that the wind velocity and/or direction were different than predicted and they had to recalculate the rest of the trip, while flying the plane and making corrections.

Meanwhile, they had to continuously check for other aircraft, their instruments, their heading and their altitude.

Piece of cake......Sure!!!

Also, they'd check "radio-bearings," change radio stations, check in with other Airport Towers if necessary, but otherwise,

AT-6 Texan – The Yellow Monster

they'd have little to do.

Heaven forbid if they got lost, oh boy! That was their greatest fear.

The first few Cross-Countries with an instructor were kinda secure, because an instructor could always jump in and save them, but once they went solo..... that was a different ball game. All of a sudden, all the problems were their own.

The plane was conveniently equipped with a "Relief-Tube", so that a pilot under "pressure" could stick his ding-a-ling in the tube and relieve his bladder. On a 3 hour flight, that could happen.

The illustrious "Choir-Director" had a different problem. The hysterical story went as follows: On his first solo-cross-country, his bowels were giving him warnings. He tried to ignore them, he considered landing someplace, he thought of making in his pants, but finally found a better solution.

He climbed out of his parachute, stashed it between the front and rear cockpits while maintaining altitude and direction. Some feat!

He then tore pages out of the flight manual that belongs in every plane, gently placed them in the seat where his chute used to be, unzipped his flight suit and wrestled it down around his knees.

All the while, he's still flying the airplane, trying to maintain altitude and course!

He relieved himself neatly on the pages in the metal seat and wiped himself, still controlling the plane. The guy had to be one hell of a pilot!

He got himself re-zipped, carefully folded his droppings in the pages of the manual, opened the canopy and started to deposit the smelly package overboard.

Slight hitch!! The moment he got his package out into the 130 M.P.H. air stream, it blew right back at him. All over himself, the cockpit, the parachute, the seat, all over!!! Lucky for him, It spared his eyes.

He smelled like a pigsty when he landed and even though he protested; *"THAT'S NOT QUITE THE WAY IT HAPPENED!"*, he didn't live it down for weeks.

One night, after night flying, Frits was called out of his room to come to the tower immediately.

There were a few differences between day- and night navigation, for one, after completing their cross-country, the new pilots-to-be had to make a few night landings for proficiency's sake. Also, when parked on the ramp, before turning off the switches, the pilot was to report to ground control; "Ramp-out". *"STALLINGS GROUND-CONTROL, THIS 427, RAMP-OUT!"* The controller would repeat; *"ROGER, 427 RAMP-OUT!"* and mark off the plane in his log.

The reason was: at night it wasn't easy to spot the parked planes on the ramp and read their numbers. Just a double-check. One thing was the same as during daytime flights.

Over each turning point, an instructor circled overhead and checked off the students on his checklist as they reported over their turning point. He would then visually verify that the plane calling in was really where he was supposed to be: turning on a new heading, right over his checkpoint.

It might go like this: "La Grange control, this is 247 over LaGrange turning to 095 for Washington, over!"

"247, this is LaGrange control, do not, repeat, DO NOT have you in sight, start a 30 degree bank to your left NOW! over!"

"Roger, this is 247, rolling left, over!"

"Roger 247, have you in sight, continue on course, over."

"Roger, this is 247, turning to 095, out!"

So at each checkpoint the instructor, with his list right in front of him, should check off a student every three minutes. If not, he'd somehow have to locate the wayward sheep and get him back on course.

This particular night, "394" was missing. Pilot: Jesu DeRavenge. The French called him "JESU"("DJAIZUU"), but the Americans just called him "JESUS".

Well, they wanted Frits to talk to him in French on all the available channels, locate him and then bring him in.

"When was he supposed to be in?"

"40 minutes ago!"

"JEEZ, no pun intended,,, how's his gas?"

"Another 30 minutes. Maybe!"

"JEEZ, again! On which mike?" *"JESU, JESU, ICI C' EST KINSTON, JESU, JESU, ICI KINSTON, RAPPELEZ-VOUS, S'IL VOUS PLAIT!"*

NOTHING!!!!

"Give me another channel, the one en-route"...Jesu, Jesu,

Ici Kinston, c'est moi, Frits. Pouvez-vouz m'entendre?" "Jesu?, Jesu?"

Silence!!!, Next channel,,,,,, same results!

"Did somebody check his ramp position? Maybe he's there!" "Yes, yes, yes. We thought of that. The plane's not there!"

"TROIS-NEUF-QUATRE, KINSTON TOWER, OVER!"

"Next channel!"

"Did anybody check his bed?"

"Whadda you think, he took his plane to bed?"

"You don't know these crazy Frenchmen!"

"As a matter of fact, I'll believe anything about these crazy Frenchies. Go check it out!"

"I'll go with you!" Jesu's instructor followed Frits down the stairs, two rungs at the time and over to the barracks on the double.

"Do you know where he lives?"

"Yeah, I do!"

They rushed into the barracks, then the room and....there he was!

Sound asleep!

They shook him; *"JESUS!"*

"Huh?" His sleepy eyes stared at them.

"JESUS..., WHERE IN THE HELL IS THE PLANE?"

"Ze plane???, Ze plane??? Oh!!!!! Ze plane!!!!! Ze plane is in ze deetch!"

"Sir, I think he left the plane in a ditch somewhere!"

"THAT SONOFABITCH!!! WHAT DITCH??"

"At ze end of ze runway."

"YOU BETTER GET YOUR ASS OUT OF BED AND SHOW US THE DEETCH!" "I'M TELLING YOU BUDDY, WAIT TILL I'M THROUGH DEETCHING YOU! . YOUR FRENCH ASS MAY END UP IN EVERY DEETCH IN CAROLINA! LET'S GO!"

Apparently, on one of his landings he had landed long and when he gave it the power for a go-around, he ran out of runway before he got her off the ground, so he ran onto the grass, cut the throttle and ended up crashing into the ditch at the end of the field. He had calmly turned off all the switches, climbed out, left his parachute, walked home and went to bed!

This might have worked at the front during the war, but in Kinston it earned him a one-way ticket back to MOTHER-FRANCE!

Things occurred and will probably continue to occur to prove that there is a HIGHER JUSTICE!

McGough found out!

One such time was a sunny day, but chilly morning in Kinston N.C.

When a pilot was in the instrument phase of his training, it meant two things; ONE, he was close to graduation, because that was the last phase of his "PRIMARY TRAINING." TWO, it was the hardest thing, he'd ever done in an airplane.

In the T-6, the rear cockpit had a cloth canopy, that fitted within the plexi-glass dome and could be pulled forward and latched. Once that was done, the student was cut off from the outside world and had to totally rely on his instruments. They would tell him if he were diving, climbing, level or turning or even if he was upside down.

What made it hardest of all, he had to learn to completely disregard his feelings and senses and absolutely believe his gauges. He might be level, yet feel like he was in a bank. He had to accept that the plane was LEVEL if the instruments said so.

Everybody struggled with this, including a young student who was sitting at the end of the runway, lining up for an "INSTRUMENT TAKE OFF," an I.T.O.

His instructor in the front seat was his "safety-pilot," ready to take over the instant something went wrong.

While the student started his "take-off-roll," John T. Mccough was taxi-ing out with one of his students. His canopy was rolled back, his fur collar up, sunglasses in place and his baseball cap squarely secured on his head underneath his "head-set". Everything was perfect on a "perfect" flying day.

McGough's student was carefully zig-zagging his way toward the beginning of the runway, while 50 feet to his right, an instrument guided T-6 raced past them in the opposite direction.

Suddenly, that T-6 was off the ground,,,, WAY OFF the ground....and BOOM,,, back on the ground, SO HARD,,,,,, the right front wheel broke off, the wing hit the dirt, causing the plane to "ground-loop" in a complete circle. The left wheel snapped off and flew way up,,,,, way, way up in the air.

Then, in a leisurely arc, the wheel came back to earth and hitMcGough smack in the head, knocking him out.

His student kept right' on taxi-ing!

Who said; "THERE AIN'T NO JUSTICE?????"

McGough just got HIS!!!! There had to be a God somewhere!

"*AEROBATICS*", the "roller-coaster" part of flight-training was the most fun!

It took more precision than Frits had anticipated, but with training, "precision" also became second nature.

The most intricate maneuver was the "*EIGHT-POINT ROLL*". It was like a "*SLOW-ROLL*" with 8 distinct sides to it.

First, you'd have to roll the plane into a 45 degree bank, but without letting the plane turn. The nose had to stay at the samepoint on the horizon.

In order to achieve that, you'd push the opposite rudder.

At that point, you'd be flying "cross-controlled."

Then you'd bank another 45 degrees, so the plane would be standing on it's side, while still flying straight out.

To keep the nose from dropping down, the top-rudder is making like the elevator, with the top-rudder-pedal all the way in. The plane naturally wanted to turn, but forward pressure on the stick kept it in line with the reference point on the horizon.

Another 45 degrees for a total of 135 and nearly inverted. The forward pressure on the stick had to increase, power had to be added, because the plane started to lose lift and speed and the pressure on the top rudder could ease off.

45 degrees more and you'd be on your back. Nose still pointed in the same direction, stick way forward to keep the nose above the horizon, rudders neutral.

45 degrees again and the whole procedure was reversed until 360 degrees had been turned and the plane was straight and level, still aiming at the same point and back to the same speed and without loss or gain of altitude. Not even a foot!

The problem usually was that after rolling 8 POINTS, they ended up "NEARLY" level and had to sneak a few more degrees in order to end up perfectly level.

Like a ninth facet on an eight-sided diamond.

That would call for the usual; "*I ONT THAINK YOU AIN'T GOT NO DARE JAM BIZNIZ IN DIS HERE FLYIN' BIZNIZ, I 'ONT THAINK!*"

"Oh Shit!" Try again.

The whole maneuver barely took ten seconds, but it could make him lose 100 feet or knock him off his course by 20 degrees in a heartbeat. That was not supposed to happen.

"*SNAP-ROLLS*" were more fun. None of that "precision" stuff.

You'd pull the plane into a "near" stall, yank the stick into your gut and to the side, while at the same time "busting" in top rudder.

With a "SNAP", the plane would turn over like a pancake on Sunday morning. It might rattle their teeth, but the plane kept right on flying.

"SPINNING" was more gentle in that monster than in the TIGER-MOTH. It seemed a little more controlled. When in a spin, he could predict how much longer it would turn in it's race to the ground. From the moment he broke the stall, by pushing forward sharply with the stick and neutralizing the rudders, it would make one more full turn before coming out and gaining speed rapidly.

Just a little throttle and pulling back on the stick got him back to straight and level flight in no time at all.

The TIGER-MOTH had more of a mind of it's own. It would come out of a spin alright, but in it's own good time and pace. The T-6 was neater.

Going up solo with the assignment; "Go wring her out!" was like going for a joy-ride in a Ferrari!

If he made stupid mistakes up there, plenty of altitude would give him a chance to recover without busting his ass.

"WHEEL"-landings, rather than "3 point landings", made the whole landing procedure much more interesting. With a "power-on" BOMBER approach, it was easier to land on the precise DIME where he wanted to be. This also helped in "cross-wind" landings or if he came in too high and "slipped" the plane all the way down final. With "power" on and touching on the front wheels first, it was much easier to correct mistakes and make it look good.

All aspects of flying seemed to become more fun and more or less "second nature", except INSTRUMENT FLYING.

To prepare the young aces for instrument flying, for I.F.R.

conditions, (Instrument Flight Rules) they had to spend long hours in "LINK-TRAINERS".

These were little wooden airplanes, mounted on a pedestal, that could rotate, bank and simulate just about any kind of flying conditions.

Charming young Lady-Instructors sat behind plastic covered desks, complete with headsets and microphones. They were in complete control of the LINK-TRAINER and the pilot inside that monster.

With exactly the same instrumentation as in the T-6, the pilot was instructed to climb, descend, turn, home into a radio-beacon or come in for landing on a G.C.A..(Ground Controlled Approach).

The class kept shrinking*!!!... ... SCARY, REAL SCARY... ..* Many fine young boys just went home. Find another way of life.

I.T.O.'s (instrument take off's) and G.C.A.'s (ground controlled approach) were the most difficult.

From the time the Approach Controller announced; "From here on in no affirmation will be required!", he talked incessantly:

"Three-two-four, you are now on your final approach to runway "zero-niner". Start your let-down "now".

You're on course and on glidepath, turn left two degrees to heading 088, your new heading is 088, you're above glidepath, increase your descent and turn to heading 086, zero-eight-six your new heading, you are back on glidepath, you're off to the left of centerline, turn right to a heading 090, you're below glidepath, decrease your rate of descent, your heading zero-nine-zero, still below glidepath, decrease your rate of descent. You are to the right of centerline, turn left to zero-eight-five, zero-eight-five your new heading, you are on glidepath, you are on centerline, you are approaching touchdown. You are on centerline, you are on glidepath. You are over touchdown, take over visually."

Meanwhile, the poor slob under the hood, with nothing in front of him but a bunch of bouncing needles, was working up a sweat. That G.C.A., a minute ago, was a near perfect one. It usually didn't work that way.

Two or three degree corrections don't even require dipping

the wings. The guy just pushed the rudder pedals a little, but that only came with experience. At first, their corrections would have the poor plane all over the sky. Way too high, way too low, ten degrees off, one way or the other, it was terrible. Just terrible!

Little did Frits realize that the G.C.A., that qualified him on instruments, one that he really wrestled with at the Marine Air Station at New Bern, was just a minor "prelude" to the G.C.A. he'd fly later at Fort Hood Texas for his Instrument Ticket. Obviously "flying" was not an Air Force privilege anymore. The Marines and Army were in on it too!

A pleasant interlude was CHRISTMAS LEAVE.

Frits' roommate, Ed Brown, from Towson MD., invited Frits home. A lot of the American Cadets invited the foreign students to spend Christmas with them and their families. It was a great gesture and of course a lot of foreign kids jumped at the chance. Frits toyed with the idea of going to Florida with some of his Dutch cronies, but the chance of spending time with a real American family in a real family setting might not present itself again. Besides that, Ed was a most pleasant and highly intelligent young man.

The folks up in TOWSON ran a "nursery" and lived in a big, three-story frame house.

On the third floor, the boys had their quarters. The boys being: Ed, Frits and Don, Ed's younger brother, a Coast Guardsman, also home on leave.

That proved to be a blast. The boys shared one common interest: "PARTY!"

Not a serious thought entered their minds, unless it was: "What time do we meet the girls?", or something equally important. A "MARINE" buddy of theirs, "Danny", made it a great quartet,, ready to raise hell.

U.S. Air Force, Dutch Air Force, Coastguard and Marines together spelled; "RAISE HELL!" and in many ways they did.

The two Brown boys and Danny had lived there all their lives and knew all the places to go and all the girls to see.

Ed was actually the only one who had a "steady" back home, a Nursing Student from Salisbury, so many times their plans included her, such as the Christmas Party and the New Year's Eve Ball.

But many times, she was not included and the guys would be out for "rowdier" times.

One such time was when Ed had lined up some girls to go dancing with. It turned out to be a disaster for Frits.

For one thing, he looked so damn young. Always had.

For another, he carried no I.D., his Passport was left on base. So, as they were making their "pick-up" stops and the car got loaded up, they made one last stop for Frits' blind date.

EVE was a nurse, a buxom blond of about 25. So far that sounded good, but when she was mashed into the car with all the others and introduced all around, she shot at Ed; " What did you get me here Ed, some kid?" MOST humiliating.

Frits now had something to prove and he was determined to make her feel differently before the night was through. WRONG! At the club, four couples seated snugly around a table, the waitress took their drink orders and asked Frits for an I.D. "ONLY FRITS!" Not Don who was younger... NO... . only Frits. He was WELL over 21, already by ONE month, but couldn't prove it.

"I'll bring you a coke!" she announced and that did it!

He got up, went 'to the bar and bought a beer from the barman, who didn't question him.

Eve had snuffed him out as if he wasn't there and Frits roamed around, dancing with other girls.

The evening was a flop.

The NEW YEAR'S EVE dance was great! The night was cloudless and bitter cold. The whole world was white and beautiful.

The other guys brought their dates to the dance and Frits latched on to a set of beautiful blond twin-sisters.

The music was "GLEN MILLER-TOMMY DORSEY" style and he danced his shoesies off. The twins were very hard to keep apart, but the one, *LISA,* seemed more interested in Frits than the other, who kept dancing regularly with a tall skinny guy. That was a nice solution, because it helped Frits to make up his mind as to whom he was going to cling to.

Cling they did! The slow rendition of "*Once In A While*" had them cuddling together so tightly, they might as well have been glued together. Lisa coo-ed; "I love your singing" and that was all the encouragement he needed.

"Once In A While, Would You Try To Give One Little Thought

To Me? Though Someone Else May Be, Nearer Your Heart?"

The pressure of her hand on his neck increased and his right hand pressed down harder on her lower back.......

"Once In A While, Would You Think Of' The Moments I Shared With You? Moments Before We Two Drifted Apart?"

The world around them was forgotten and they were only aware of each other and the music.

Their lips brushed ever so briefly, then returned for more of the same and before long, they were passionately kissing on the dance floor, while moving a little with the music. VERY little!

Frits was thinking to himself: "I was right, when I was telling Ed that girls love it when you get a hard-on on the dance floor. They love the thought that their femininity can arouse a man like that!"

Ed had disagreed and put a jockstrap on. "NUTS!"

Lisa and Frits couldn't get enough of pressing their contours into one another. Too bad that fast dances interrupted their mood every so often, but as the evening wore on and the endless supply of Champagne kept flowing, they didn't give a damn, what the band played, they just shuffled slowly across the floor, glued together.

After midnight, the Twins, Skinny and Frits were hustled off to a private party with a lot of drinks and good wishes and off again to another party with more of the same.

All that time, the young lovers clung to one another, in and out of backseats, in and out of living rooms, but at no time did they get the chance to consummate their newfound love.

Then, suddenly, in the middle of the night, the girls got out of the car and were home. They disappeared by themselves into the darkened house and that was it. Over... gone.... alone.... their Daddy would have killed them and everyone else, if they had brought them into the house, so.....all of a sudden.., no more love!

The nice, unknown driver drove him to the bus station in Baltimore and after a long wait, he boarded 'a bus to Towson and after a half hour walk in the bone chilling cold, he arrived at the Brown's, sobered and frozen, frustrated and exhausted, mad and disappointed.

He didn't even know her last name...., at least he didn't remember....... It was six in the morning.

His hangover was the worst one of his life.

His head throbbed so bad, he thought his brains would explode.

To make it worse, he had to be up, dressed and shaven, to have breakfast with the Browns, celebrating the New Year. He must have looked real bad and cross-eyed, because several antidotes were suggested: *"LOTS OF MILK!"*, *"LOTS OF ORANGE JUICE!"* Nobody seemed to appreciate a good hang-over, otherwise they would have put him back to bed.

They had a "lunch and Rose Bowl" date with Danny's parents in their Baltimore apartment, across from the BENDIX plant, where Danny's father worked.

The living room was cramped, the T.V. too loud, the company too funny and the headache too miserable.

Dan Senior, having sympathy, fixed a "Scotch and Soda", but Frits couldn't down that for nothing. It tasted just terrible. When he asked for a beer, the headache started to diminish considerably. Three beers later, his appetite came back and after a six-pack, the Football game even started to make sense. With very patient explanations from Mr. Dan, he finally got somewhat of a grasp of the game, but not enough to understand their screaming and enthusiasm for the spectacle.

By game's end and after some bar-hopping, his spirits, his head and his voice were back to normal and they enjoyed a late last fling in good old Baltimore!

HAPPY 1953

Things were starting to wind down at Stallings Air Base.

The boys were now First Class Men. Frits had been brought in for an interview concerning a "Cadet-Officers" position. The foreign cadets required a "LIAISON-OFFICER" with the rank of Cadet Captain.

When he was informed that he was one of the front runners for the position, he was insulted.

"Who else could they possibly consider?" " A Dane? Really, all he speaks is Danish and English, he's got no way of communicating with the rest!"

"You really consider that an option? Well, shove it! I don't want it!"

He obtained the rank of Cadet-Sergeant, along with Loutje. It was the lowest rank available. The foreign cadets considered him their only "liaison" anyway, regardless of rank.

A "PROM" was organized for graduation, a nice, dressy affair and one of the Goldsboro friends, a Dodge dealer, lent Frits a Coupe for the occasion.

The man never even asked him for a license. He probably had never met a 21 year old before, who didn't have one.

Frits had the car for 4 days and it put him in a position to properly escort a Goldsboro Blonde to the dance.

It was swell, but it was sad. Everybody was being scattered all over creation and only "Loutje, Wim" and a few Frenchmen were transferred to the same base in Texas for "BASIC TRAINING". All the rest went elsewhere.

The small group was placed on a DC-3 and flown to "College Station", Texas. From the air, it didn't look like much.

CHAPTER VIII
BRYAN AIR FORCE BASE, BRYAN, TEXAS.

First impressions can fool you!

From the air, everything seemed rust brown and dull.

The town of COLLEGE STATION... . dull, Texas A&M... DULL, BRYAN, the base... dull... THE LAND AROUND... DULL!

The welcome was just the contrary: WARM... WARM... . WARM!

Captain and Mrs. ROXS, smiling broadly, welcomed them. Sergeant PIETER GROOT hustled them off in his car and they were made to feel at home immediately!

Capt. Roxs was the Dutch Liaison Officer for all the Dutch boys in the U.S. and he shared an office with a Belgian Major, who was his counterpart for the Belgian Air Force.

Pieter was a washed-out cadet, who had managed to stay behind as Staff-Secretary and in Roxs' absences, kinda ran the show.

It seemed like a nice set-up.

The Belgian Major, it turned out, was born less than 20 miles from where Frits was born and he enjoyed reminiscing with the young Dutchmen in a mixture of French and broken Flemish.

Pieter chauffeured the boys around and saw to it that they were properly booked-in and comfortably situated.

There were a few familiar faces; Pierre Gilliam and Henk Benninga had been upperclassmen at Kinston and Jan Mante had been on the boat with them.

There were a total of 15 Dutchmen spread over four classes. It made them feel at home instantly.

Of course, they were "Fourth Classmen" all over again, but it didn't involve as much chickenshit as they had suffered through in *"PRIMARY!"*

Classes began immediately. Engines: *"DIFFERENT"*, Instruments: *"DIFFERENT!"*, Radio's: *"DIFFERENT!"*, they even had U.H.F.!

" Flight procedures" and the Type of Aircraft were also brand new and they learned that they would spend 65 hours on a T-28, and then onto the "JET", the T-33 "SHOOTING STAR" for an additional 65 hours.

The T-28 was a "NOSEWHEEL" type plane. Their first.

It had a powerful engine and huge flaps, they were lovingly referred to as "church doors." In order to climb unto the wing, they had to make two steps up into the lowered flaps.

No more "zig-zagging". When they sat up in the plane, they just looked over the nose of the plane and rode off with a clear view over the top of the engine.

From the first ride, Frits was in love with the plane.

WHAT A LIVING DOLL, compared to that Yellow Monster!

His new instructor was First Lieutenant Richard Rittiger. Rittiger, like most instructors, was back from a few tours in Korea, mostly on F- 80's, the "MOTHER" Shooting Star. Some had flown P-51's, the old W.W.II MUSTANG and some on the F-84 THUNDERJET.

Lt. Rittiger was a gruff talking, likable young man with the worst cursing habits Frits had ever heard in his life.

He would even split up a word, in order to get a curse word in between! A T-28 would become: *"A GODDAMN T-FUCKING TWENTY-EIGHT!"* Frits was not bashful by any means, but he did make up his mind, he'd never swear like that!

Rittiger was a patient, easy going teacher. Very unlike McGough had been and he had Frits enthused about the plane in no time.

With all it's power and light controls, it did aerobatics so easily, it was like child's play. He could have done "cloverleafs

Karel Ten Hove and Frits with the T-28

and cuban eights" all day long.

A "cloverleaf" is a nice, easy-going, continuous roller coaster maneuver. They'd find a perfect perpendicular highway crossing on the ground below.

From a few miles out, they'd start a slight, power-on dive, straight down the highway. Shortly after crossing the perpendicular road, they'd pull up as in a normal loop, but as they approached the vertical position, they'd start a 90 degree turn and by the time they were upside down, they'd be looking down the crossroad. They would then pull through as if completing a normal loop, diving straight down to earth with the engine screaming at the bottom of the loop and then up, up, up, again toward the heavens and again, when standing on their tail, they'd start to roll again and end up upside down again, heading a 180 degrees from the original direction.

In other words: they'd pull 4 continuous loops, turning 90 degrees when on top and back into the original position and direction as when they started.

With that T-28, he could have done that for hours on end.

A "Cuban 8" involved no change of direction. Flying over a straight line, he'd start a normal loop, but when coming over the top, instead of completing a full circle, they'd roll over to a "wing-level" diving position and start into another loop. Repeat and repeat and repeat until he'd tire of it, all the while making sure that he stayed over his straight line on the ground.

They actually "wrote" a vertical 8 in the sky, whence the name.

This T-28 was one hell of a flying machine.

Coming in for landing, they'd come in so close to the runway on downwind, they were nearly overhead. Instead of the old "90 degrees to base-leg" and 90 degrees more to final, in that baby, with the "church doors" fully extended, it became more of "180 degree dive-bombing pass!"

That was exciting! Sometimes, it got too exciting!

Once, on a solo flight, oil shot from under his engine's cowling flaps, all over his windshield.

His reaction was immediate; Retard throttle, check oil pressure, head for home, check for emergency landing field and inform the tower.

"Bryan tower,this is 488, no "may-day" yet, just an oil leak, pressure steady, 15 minutes or less from touchdown. Windshield splashed with oil, forward visibility poor,over!."

"488, Bring her straight in to runway 0-2, we'll clear the pattern, maintain enough altitude for possible bail-out"

"488, Roger!, Have the field in sight, no forward visibility,

I closed the cowl plaps, but the engine temp went up, so I re opened. Oil-pressure still o.k., over!"

"488, You're number one on final, clear to land"

"Roger, clear to land, 488!"

While he was not able to see straight forward, by slipping the plane a little to the left or to the right, he could see enough past the windshield to stay lined up with the runway and once down, he simply leaned out far enough to stay lined up with the runway's edge.

"488,Make first turn-off to the right and shut the engine!, Over!"

"Roger,488 turning off, changing to ground frequency, out!"

It had all worked out like a charm. Luckily, he had not blown out all of his oil, because a "dead" engine would have made a lot of difference. As long as the engine put out power like it did, it was like normal flying to him. No sweat!

The T-28 was equipped with a "relief-tube" as well. Where it differed from the one on the T-6, was in the drainage.

The T-28 did NOT drain into the air, but into a tank that had to be drained regularly on the ramp. Frits had never used one yet and had never given it any thought either.

Till one day....... while hanging upside down in some maneuver, he did not maintain positive "G's", but somehow held up the inverted nose of the plane with negative "G's".

In other words, he was screwing up!

He knew it immediately of course, cause he was hanging in his shoulder straps instead of being pressed down in his seat. He also knew it because of all the cigarette butts and pencil stubs that fell upwards from the floor, passing his face and landing in the top of the canopy over his head.

Then..., with a swishing sound......the relief tube emptied out into his lap and ran right down his upside-down body to his face.

By way of his hair, it fell into the bubbletop of the canopy with the butts and the stubs.

He rolled the plane back to the right side-up position so fast, that the fermented pee did not run by the side of the canopy down to the floor, instead, it landed smack back on top of his head and ran back down his chest into his lap.

He cursed like a dock worker and out of all the stupid

things to do, he licked his lips. He spit as fast as he tasted the vile mess and it landed right on the altimeter.

Heaven knows, how many people had peed in the damn thing and how long it had been since it was emptied out. In the boiling 100 degree temperatures, that human sewage had cooked into the stinkenest garbage he had ever smelled and now he had it all over him.

He headed for home, slowing down the airplane, so he could open the canopy and at least breathe some fresh air.

When he landed, he was fuming, steaming and stinking. He was gonna burn some crew chief's ass for sure. In the two years he had been in the service, he had never reported anybody yet, but today was the day.

Unfortunately, the crew chiefs laughed hysterically, the pilots laughed just as loud and wouldn't even let him into the crew room, because of the way he stunk, and his instructor couldn't stop laughing either.

There was nothing left for him to do, but walk to the showers and strip while the water ran down his flight suit, his gloves and into his boots.

He threw away his charts, clipboard and notepads. Everything stunk to high heaven.

Back in the crew room, an hour later, word had gotten around and the ribbing he got was just endless.

In the cadet club, later on, the other flight members were told about it and by then he started to see the humor in it too and he decided to relate the whole gory story in detail, creating more hysterics. It was THE topic of the day.

One of the new aspects of the training was "FORMATION FLYING". At first with just two planes, each of them manned by a student and an instructor.

To start off with, it was a scary experience, flying that close to another plane, while making only minute adjustments with the throttle and the stick. The trick was to stay right in position with the lead plane, whether it was flying "straight and level", "climbing or gliding" or standing on it's side, but like all other phases of the training so far, it got easier with practice.

Pretty soon, Rittiger would be up there with four of his students

He'd be in the lead, flying from the back seat with a student in front and the other three planes would be flying off his wing, or wings.

There were different formations;

Standard was: "lead" up front, "#-2" on his left wing, #-3" on his right wing and "#-4" on "#-3"'s right wing.

Coming in for landing, they'd shift into an "echelon" formation. The "lead" up front of course and 2, 3 and 4, off his right' wing, just slightly behind each other.

A "BOX" would have #-2 on his left wing, #-3 on his right and #-4 below and behind him, thus forming a box.

They felt they were getting to be pretty good. Maybe too good.

So good in fact, it earned Frits his first "pink-slip", a failing flight!

If a pilot was sitting off the leader's right wing and was signaled to switch to the left side, the procedure was as follows: Pull back on the throttle in order to slide back, give it throttle again to stop the slide, when clear of the "leader's" tail, when absolutely behind the lead aircraft, bank a little and slide over to the other side, give it more throttle to slip forward into position and retard throttle in order not to fly past him.

Simple enough? Nuttin' to it, right?

Rittiger (left) T-28, Nicolas and Tita (on the wing)

Well, on one flight, Frits moved from his right slot to his left slot so fast and so nicely, that he was in position in just seconds.

Great? Not so great!

"*#-2, get the fuck on home. now!..Now!*" Man, Rittiger was screaming!

On the ground; "You passed that fucking propeller so close to my goddamn ass, I thought you'd cut my fucking tail off! You never dropped back far enough to clear it, you ignorant son-of-a-bitch. you could have gotten my goddamn ass busted and I like this fucking life and intend to stay around a goddamn bit longer.

I survived two hundred flights in Korea and here I'm about to get my ass busted by some stupid Dutchman and you have a goddamn pink-slip up your ass, buddy, and a next goddamn time you pull that shit, you're outta here, you hear?"

"But Sir, I was clear of your tail, sir!"

Rittiger kept on screaming: *"Not in my goddamn book, you weren't. and that's what counts, goddamnit!*

Pink slip, failing ride!"

During their "hanging-around" time, when the instructor was up in the air with someone else and he had no "solo" scheduled, he could stand and look at the "JETS" for hours.

Taxi-ing in or out, taking off or landing, coming in on the "break" in formation, he could never get enough.

"If I can only get to solo a T-33 once, I'll be happy", he mused to himself. "If I then "wash-out", I won't complain. They can ship me back to Holland, but I'll know, that I solo-ed a "jet" once. Just once, baby, JUST ONCE!"

He'd be talking to himself or with his buddies as they watched the continuous coming and going of the sleek aircraft in the hot Texas sun...."JUST ONCE!"

His chance came faster than he expected.

After finishing 65 hours on the T-28, they were issued *"helmets"*. White plastic helmets, complete with earphones built in and an oxygen mask attached that included a microphone! How classy!. On top of the helmet was a *"visor"*,a sunglass type of contraption, that slid up or down as needed. Now they could climb to 40.000 feet and survive! Each pilot could go to the paint shop and have his own design painted on. Talk about inovation. Frits designed a blue helmet with a bright yellow F, backed against an inverted F. It was a design that dated back to his childhood when he imagined being a Viking Warrior.

Each instructor ideally had 4 students, sometimes five. As some washed-out, it could become two or three.

John Rittiger had a perfect four. Not that his students were so perfect, OH NO!. The number was perfect.

Joe Testa was already a 2nd Lieutenant. He was a rarity. He was the only Jew Frits knew in the entire flying system. He was from Philadelphia and showed a keen interest in Frits' wartime experiences under German occupation.

He also corrected Frits' English whenever necessary and Frits found that most helpful.

When the guys were asked to write a brief autobiography, Frits wrote amongst other things; *"I GOT A MOTHER, TWO BROTHERS AND THREE SISTERS!"*

Joe corrected him; *"YOU DON'T "GOT", YOU "HAVE!" A MOTHER, ETC."*

"Yes, but everybody.."

"I know, I know, NEARLY everybody MEANS to say; *"I'VE GOT!"*, but they say; *"I GOT!* It's still wrong. You need to say; *"I HAVE..!"*

"Thanks, Joe!"

"CHUCK" was the typical, tall, good looking, all American boy. A very pleasant, studious kind of guy. All his ducks in a row, girl at home, future planned out and well on course.

Nice human being with a ready smile.

"PIERRE", a quiet Frenchman, when in the company of the instructor, a "wild-man" when surrounded by other Frenchmen. He had a better grasp of the English language than most and he appeared to be a good pilot.

And then there was Frits.

After work, a sort of routine was established: Chow, then the movies in the base theater, highlighted by the newsreels and some Loony Tunes cartoons, then to the cadet club and "flying stories" or other unusual or exciting events that happened that day in the sky. They could shoot the breeze and tell jokes about that forever.

Many a time, they'd hit the town in Pieter's Studebaker for a few beers and some laughs.

Since Pieter had a "PERMANENT STAFF" sticker on his car, and the guys were always in civvies, they were never checked at the gate and cadets, who had no business being off base during the week, were happily jiving around town anyway.

The only thing wrong with that schedule was there was no time to study! Of course, that type of thing would catch up with somebody, sometime. And...IT DID!

Each morning over coffee, Lt. Rittiger would go around the circle: "RADIO FAILURE", at that point the procedure should just run off his tongue!

"ELECTRICAL FAILURE!", same thing. Lots to remember. For instance, FLAPS were "hydraulically operated", but "electrically actuated", so with an "electrical failure", you'd have NO flaps. Also NO radio, NO fuel pumps, NO vacuum pumps, most instruments "OUT" or "tumbling!". Interesting problem.

"HYDRAULIC FAILURE!", the "dive brakes" wouldn't work, the "landing gear" wouldn't "lower and lock", not funny at all.

"FLAME-OUT!", meant that the engine quit. "How to restart it?"

Knowing how to react instinctively to the potential emergencies was of the utmost importance. Knowing the procedures was paramount to being able to react properly. So...each morning, they'd go through this routine of rehashing "emergency procedures", before going up for instruction.

The norm was three to five dual-flights and the student should be ready for solo.

Frits had been up twice. Total hours; 2 hrs 48 min. He fully expected to solo after the next dual ride with Rittiger.

So did the other students. But...there they were again...

"Chuck, Radio Failure!" "Zip, zip, zip," he knew it inside out.

"Joe, Hydraulic Failure!" "Zip, zip, zip.!" he just spit it out.

No hesitation, whatsoever.

"Frits, Flame-out!"

"Uh...I 'think...uh.., stopcock throttle, .uh. .pull—up nose. ..uh... .drain . .uh. .tailpipe".

"Pierre, Flame-out!" "Zip, zip, zip.!" With a French accent.

"Joe, Power-on Stall!" "Zip, zip, zip!"

"Chuck, Electrical Failure!" "Zip, zip, zip."

"Frits, Oxygen Failure!"

"Uh. . I think.. .uh."

Rittiger started shouting: *"YOU LAZY FUCKING SONOFABITCH, GET YOUR GODDAMN GEAR, HAND ME YOUR FUCKING LOGBOOK AND MEET ME AT THE GODDAMN DISPATCHER.,YOU DUMB FUCKING DUTCHMAN!"*

Frits, thoroughly startled by the verbal explosion, handed over his logbook and walked out to gather up his helmet and parachute. He met Rittiger at the dispatcher's, where he was handed a small clipboard.

"YOU GOT FOUR-THIRTY-FUCKING-THREE, FORRER, . NOW STRAP YOUR FUCKING ASS TO THAT GODDAMN PLANE AND GO AND DO ME A FUCKING FAVOR! GO AND BUST YOUR ASS. KILL YOUR GODDAMN SELF. DON'T COME BACK!"

He was shouting again!!! It was embarrassing!

Totally baffled and shook-up, Frits walked down the flight line, looking for 433. "Oh my God, I should have studied more!" He talked to himself as he was walking, "Hotdamn, I hope I can remember all that shit!. ...Well... I'm gonna fly it anyway, so I might as well quit worrying about, it!"

After locating the plane, he dumped his chute in the front seat, checked the cockpit briefly, tied the shoulder straps in the rear seat and started the exterior check, with the help of his checklist.

He made up his mind *"I'LL SHOW RITTIGER!"*, so he checked everything very deliberately, making sure he didn't make a single mistake. He would prove himself; *"JUST WAIT RITTIGER, JUST WAIT!"*

The plane had no tip tanks, so Rittiger would not have to wait very long. He could barely stay up for 35 minutes like this.

Satisfied with the exterior of the plane and having removed all the safety-pins and covers, he settled into the cockpit. After strapping himself in properly, he plugged in his oxygen and radio connectors and started his cockpit checklist. Nothing was overlooked. He couldn't afford to. One oversight might kill him. He wouldn't give Rittiger that pleasure.

When ready to start, the crew chief signaled "clear" and the engine roared to life.

"Check all instruments, turn on all necessary switches!" he talked to himself again, like on that first solo nearly a year ago,

"Now radio-check". He was talking to himself again.*"BRYAN GROUND, THIS IS 4-3-3, OVER!"*

"4-3-3, THIS IS GROUND, GO AHEAD!"

"BRYAN GROUND, THIS IS 4-3-3, RAMP "E", READY TO TAXI FOR THE FIRST SOLO, OVER!" They had to communicate to the tower, that they were on their "1st", "2nd" or "3rd" solo.

"ROGER, 4-3-3, TAXI TO RUNWAY 1-2, OVER!"

"ONE-TWO, ROGER!"

At the beginning of the runway, where a remote unit controlled the take-offs and landings, he went over the "pre-take-off"' checklist with the controller, as required on the first three solo flights.

After getting a "green" light, he lined up for take-off. Over to "TOWER" frequency.

"BRYAN TOWER, THIS IS 4-3-3, READY FOR TAKE OFF!"

"4-3-3, CLEARED FOR TAKE OFF, OVER!"

"ROGER, 4-3-3, CLEARED AND ROLLING!"

He checked once more with his right hand if the canopy was locked, while shoving the throttle forward with his left hand.

As the RPM's raced to 100%, the plane shot forward, faster than usual. Without the 500 gallons in the tip tanks and the weight of the instructor, the plane accelerated like a bullet.

Snow started blowing in from underneath, where the air conditioning was compacting the humid Texas air.

Before he was halfway down the runway, the back pressure on the stick had him flying. Holding the nose at a slight climbing attitude, he picked up the gear and started "milking up" his flaps. By keeping the plane fairly low over the runway, he reached "climbing speed" by the time he was over the end and he pulled it up sharply. BOY...did that baby climb without tip tanks and Rittiger. "WOW!.. This was flying!!

He broke out of the pattern and within a few minutes, climbing 4,000 feet a minute, he was at "altitude".

After making sure that he was "clear" up there, he looped her, rolled her, made a "split-S" and turned a huge "barrel-roll".

"WOW-WEE!" This was the ultimate!!!, *"BRING ON THE RUSKIES, BRING ON THE MIG-15'S, I'M READY." "WHERE ARE YOU? YOU CHICKEN BASTARDS?"*

Fortunately, there where no MIG-15's, because he didn't have enough fuel to chase a FLY, let alone a MIG.

He headed back, 500 MPH, and down to 1200 feet, slowing

down as he approached the base.

"BRYAN TOWER, THIS IS 4-3-3 OVER I.P., OVER!" Slowing to 400 MPH.

"4-3-3, THIS IS BRYAN TOWER, YOU'RE CLEARED INTO THE PATTERN. YOU'RE NUMBER 4, OVER!"

"ROGER, THIS IS 4-3-3, HAVE THEM IN SIGHT!"

There were two T-28's and one T-33 ahead of him, the formation of two T-28's about to turn final.

Speed down to 200 MPH, *"BRYAN TOWER, 4-3-3 TURNING DOWNWIND."*

"4-3-3, YOUR # 2."

"ROGER, # 2, THIS IS 4-3-3"

Now through his pre-landing check!

Altitude,,, speed,,, distance out from the runway,,, FINE!,,, Fuel,, FINE!,,, slow her down, 10 degrees of flaps,,, instruments ,,,FINE Gear down,,,, locked!. ' '

"BRYAN TOWER, 4-3-3, TURNING BASE FOR TOUCH AND GO!"

"ROGER 4-3-3, CLEAR TO LAND!"

"ROGER, CLEAR TO LAND, THIS IS 4-3-3."

Baseleg, 20 degrees of flaps, check height,,, distance,,,, instruments,,,,, o.k.,,, turning final,,,, good,,, sitting pretty,,, full flaps,,,, close throttle,,,, start round-out,, This is going to be nice,,,,,, *"OH SHIT! POWER,,, POWER!!!* The damn plane wanted to settle down on the overrun, short of the runway,,,,,, The damn thing didn't float at all without tip tanks,,,,

Creepingly slow, it seemed like hours, the RPM's started to rev up and when they finally did, he had to cut the power immediately again. But, the boost was enough to help the plane touch down on the runway. BARELY, but on the runway.

With his right hand he held the stick pulled in his gut, while with the left hand he flipped the flap switch UP and shoved the power forward all the way.

"BRYAN TOWER, 4-3-3, GOING AROUND!"

"4-3-3, CLEARED FOR TAKE OFF."

Stick all the way back, power building up, the front wheel never touched the ground.

"MY, MY, did he feel good about that one! That would show Rittiger!! That was the way to do it!

"FUEL?" "alright!" One more landing, maybe two. Gear up! "Atta baby, What a ride!" "What a ride!"

Twice more, he came around the pattern, doing everything just the same, except on final, he didn't quite close the throttle so soon and it worked out beautifully.

On the last pass: *"BRYAN TOWER, 4-3-3, FULL STOP!'*

"ROGER 4-3-3, CLEAR TO LAND!"

When he pulled into his parking slot on the ramp and stop-cocked the throttle, Rittiger appeared on the ladder immediately!

"YOU FUCKING BASTARD! YOU GLORIFIED SONOFABITCH! I KNEW YOU COULD DO IT!!! YOU WERE FUCKING GREAT, YOU GODDAMN DUTCH PISS-ANT!!! I TOLD THEM, I FUCKING TOLD THEM,,, MAN, YOU WERE FUCKING GREAT!!!"

Frits was more flabbergasted now, than he was a half hour ago when he was sent out to kill himself.

But it was nicer this way. That was for sure!

At the de-briefing over coffee, everyone hanging on to his every word, Rittiger interrupted; "I DIDN'T FUCKING KNOW THAT THAT GODDAMN PLANE HAD NO FUCKING TIP-TANKS. GODDAMN, YOU KNOW FUCKING WELL THAT I WOULD HAVE GODDAMN TOLD YOU THAT THE BITCH WANTS TO SETTLE IN FASTER. I DIDN'T FUCKING KNOW!"

Well, that called for a few extra beers that night. As a matter of fact, he ended up having a few too many. He must have told and re-told his "first Jet solo" story ten times, if not more. It would turn out to be one of the most exciting flights of his life.

A few weeks later, while shooting pool, he developed stomach cramps and he went to the JOHN, trying to "drop" it.

After several attempts, the pain wouldn't go away and by nine he went to bed. VERY UNCHARACTERISTIC!

Nothing tasted right and his belly ached as if he had eaten too many stolen green apples, like when he was a kid.

Burping, throwing up, sitting on the john, nothing helped. It was more his belly than his stomach actually.

Finally, he fell asleep, but was woken by a jolt of pain at about 11. Hot damn, it hurt!

He got into his uniform, forgot his cap and walked to the infirmary. The Sergeant on duty was not very receptive.

"The Doctor is busy at the moment, how can I help?"

"You can help me by getting the Doctor, please"

"Who are you and what for?"

"I'm Cadet Forrer and I have "Appendicitis", so please get the Doctor."

"How'd you get here?"

"I walked and I'm in pain, so please???"

The Sergeant laughed; "If you had appendicitis, you wouldn't have walked over here, pal! "

The Sergeant thought he was funny! "Let me give you some A.P.C. tablets, go back to bed and if you still don't feel right in the morning, come back!"

"May I use that phone?"

"Hold it, hold it! Why do wanna use that phone?"

"I'm gonna call the O.D. and my Liaison Officer, so they can get me a Doctor!" '

"Now hold it a minute! Hold it! Sit down!"

The Sergeant disappeared and was back on the scene in two minutes with a Captain who stank like "Scotch".

"Hi soldier. Stand up a minute and tell me what hurts and where!"

The Doctor touched where Frits pointed, "Ouch!"

"Yes, that's it Doc, right there!"

"Come in here, take off your pants and lie down, I'll be right back."

The Doctor reappeared with a very young blond, Nordic type of a man.

"This is Lt. Jensen. He's our surgeon, just relax!"

Lt. Jensen tapped his belly, which seemed unusually tight.

"Tell me when I tap, if it hurts more or less."

"Less,,, less,, more,,, more,,, ouch,,, ouch,,, less,,, less,,"

"Gloves, please!"

A nurse had come in and she handed him a pair of gloves.

At least she didn't smell like Scotch, like the two Doctors, she smelled of Bourbon.

The Doctor, gloves and all, ran a finger up his butt.

"Oops,,, ouch,,, "

"O.K., marker!" " Frits, more or less again, O.K.?"

"O.K.!"

"Less,,, less,,, more,,, MOST" he shouted out! "Yea, oohh, oooh!"

"O.K., good show, prepare for surgery!"

Everyone disappeared until the sergeant came in with a gown.

"You need to pee?"

"I don't know."

"Go in there and try!"

Back out, the Sergeant ordered, "Take all of your clothes off your watch too! Put this on I'll take care of your clothes and uh, cadet, uh,,, good luck!"

The nurse, a First Lieutenant, came back in to shave him.

"Hi, I'm Ann!"

It was a most embarrassing moment for him.

Just when he would like to make a manly impression, his dingaling had shriveled as if he had been sitting in ice water and she gingerly manipulated it, while shaving all around it.

He felt humiliated to the core!

By 1 AM, when surgery was scheduled, he was ready. Dr. Jensen, in surgical gear and glasses, was only recognizable by the very blond hair on his forearms.

"Do you want local anesthesia Frits, or shall we put you under?"

"NO, NO,,, local please. Can you set that mirror up there so I can see it?"

Doc Jensen hesitated a second or two and said; " O.K., one condition though. You make one wrong move and we put you out, O.K.?"

"O.K.!"

The Doc set the mirror and Frits watched him cut a neat little square in the white cloth that covered him.

He gave him two injections in his belly and after a while, he started pricking him with a needle.

"Can you feel that?"

"Yes!"

"Just a little longer!"

A few minutes later: "Can you feel this?"

"No!"

"This?"

"No!"

"This?"

"Yeah!"

The needle pricks were up to his stomach by then.

"OK Ladies!" "Scalpel!"

He cut a swift 2 to 21/2 inch gash in the skin, but Frits felt nothing.

Dainty fingers put clamps on the skin and pulled it apart, while other fingers were dabbing blood.

With another knife, Jensen cut further into, what seemed like, a fair sized steak.

Frits hadn't realized that so much meat lined his belly, it must have been at least an inch thick.

More cuts, more blood dabbed and...."*THERE IT IS*!"

"IT" nearly popped up! A little red pinkie was sticking out!

"That's the culprit" said the masked voice and at the same time he put his fingers under it and pulled it up.

Frits could feel it chugging at his stomach and coughed involuntarily.

"Once more and I turn the mirror!", the mask warned.

Many busy hands kept him from seeing exactly what was going on down there, but suddenly, Dr. Jensen showed him a jar with clear fluid and an ugly red worm in it.

"Can you believe Frits, that that little thing can hurt like that? We're going to close you up now!"

Frits fell asleep.

When he woke up, it was nearly daylight and he was in a ward. He had to pee very badly! He tried to get up and that's when he was reminded of the operation, he was banded up securely. He ran his fingers over it and found that a big bandage circled his abdomen from his navel to his dingaling.

A male negro attendant noticed his movements and shot over;

"You awright?"

"Yea, I need to pee though!"

"O.K., easy, you can get up!"

He helped him up and out of bed and steered him to the bathroom. He couldn't pee a drop, yet he felt, like he had to go in the worst way.

Then it dawned on him that the anesthesia down there hadn't worn off as yet and he actually had no feeling there at all.

Back to bed. The Doctor on duty, the Captain, still on duty, came around to check and Frits tried to convince him that he

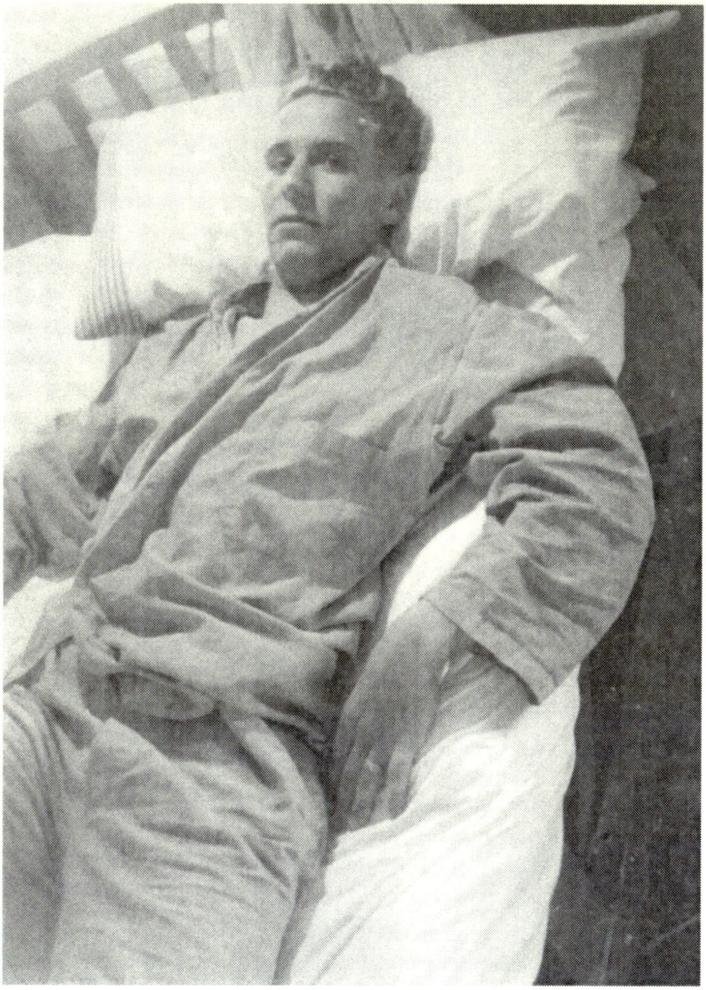

After appendicitis – Pobrecito!

was ready to go back on duty, but the Captain's answer was; "Ask me again at twelve!"

He must have known something!

By twelve, it hurt so badly, that Frits was glad to stay!

He had interesting company. Joop Tichelaar, an upper classman from Holland was over in the next bed with the same problem, but worse. His had ruptured and he was already in there for three days, with a week more to go.

Two days later Frits was discharged though, but was not to pull any "G's". He could only fly straight and level.

The best thing was though: "NO DRILL, NO MARCHING AND NO GYM!"

While the other cadets lined up and marched to and from the flight line and to and from classes, Frits leisurely meandered in the same directions, waving at his buddies.

In the air, it was a good time to practice instrument flying under the hood.

CHAPTER IX
THE EYES OF TEXAS ARE UPON YOU

Social life in Texas was different.

Lots of parties, lots of laughter, lots of beer but no girls!

Someone even suggested going to church on Sunday morning to meet girls! Heaven forbid! Heaven did forbid, because it never happened.

A new class dumped a fresh supply of Dutchman into Bryan and it added some intersting characters.

David Knibbe, who for some biblical reason was always summoned to get the beer: *"Wine, David, Wine!"* According to someone, in the Bible, David always brought the *wine*. In Bryan, David brought the *beer!*

On Friday and Saturday night, "PABST BLUE RIBBON" would show boxing movies in the Cadet Club for an hour. During that hour, while the PABST BLUE CHOO CHOO would do it's thing: *"What'll You Have? Pabst Blue Ribbon!"* the Pabst beer was free and David would do his thing: lug beer!

The rest would whoop and holler at the boxers on the screen while building a pyramid of beercans on the table.

David was from Amsterdam and so was Karel ten Hove. He talked like a real Amsterdammer, had the extremely powerful build of a weightlifter and played guitar.

Arie de Jong, also from the big city, but in contrast to Karel, he was very blond and thin, but also played a mean guitar. He had the keenest sense of humor and his rendition of *"Don't The*

Moon Look Lonesome Shinin'On Through The Trees!" was very unique and funny. When Karel and Arie got their guitars out, the singing would go on for hours on end.

Harm Appelo was from Drente, a Province in the North-East of Holland and was always treated like a farmboy by the city slickers. They picked on him unmercifully, but with his stubborn good nature, he just shrugged it off

Joop Jonkers, a tall, gangly Amsterdammer, offset some of the wildness by being a gentleman and friendly at all times.

Hans Ackerman was a big, bruising kind of fellow with a thick Indonesian accent, giving away the place of his birth and early childhood.

The Dutch group as a whole could consume a lot of beer and raise a lot of hell. Sometimes, that worked great, other times it didn't. One of the times that it did work out O.K. was a Saturday afternoon at the clay pits. It was a man-made swimming hole, not far from the base with cool, clear water.

Pieter had his Studebaker and Jan Mante had bought an old Chevy, so that provided transportation for at least twenty guys!

On the way over they bought bottles of "Muscatel" wine and sang along with Perry Como's *"Catch A Falling Star And Put It In Your Pocket*!", Nat Cole's *"I Used To Walk With You, Along The Avenue!"*, "Joni James' *"Whenever We Kiss, I Worry And Wonder*!" and a favorite of skinny Arie de Jong, "ARI-ARI", Sinatra with Billy May; *"My Lean Baby, She's So Slim, A Broomstick's Wider, But Half As Trim!"* And Jo Stafford's *"Fly The Ocean In A Silver Plane"*

Their voices might not have been the greatest, but their volume and enthusiasm made the cars shake!

At the clay pit, there was something for everybody. Shade and sunshine, easy sloping access to the water and a steep cliff for the divers.

On one side, the cliff rose straight up out of the water for about 25 feet, then a ten to twelve foot wide ledge and then straight up again for another 15 feet.

Diving off the ledge was interesting, diving off the top was close to dangerous! From some 40 feet up, it was a good high dive in anybody's book, but having to clear the ledge was something different

Lots of beer and enthusiasm lead to. "DARE" and "DOUBLE-DARE" and finally, Frits and Kareltje were going to do it.

Singing at the claypits

Spectators below in the water, up on the ledge and on top of the cliff, fully expected them to splatter their bodies all to hell on the ledge.

Frits drew first!!! Diving down was not a problem at all, diving forward far enough to clear the ledge was the true objective. Of course, the whole routine was stupid, but cheap Muscatel wine (65 cents a bottle) befuddled their minds enough that "I DARE YOU!" became a challenge that couldn't be resisted.

Getting a good foothold was most important. A slip would spell: "HEAVEN BOUND!"

Way up on the rim, he got a good foot hold, bent down, contracting himself like a spring, dove forward with all his strength, sailed down in a swan dive of sorts and hit the water rather well.

Karel, maybe more confidence in his powerful legs, made a classic swan dive as if the ledge was no consideration at all.

Pictures later showed that Karel's style was indeed superior, but their clearance at the ledge was just a matter of inches, mere inches.

Later on, more "*DARE YOU*"! They climbed a tree in order to gain an additional six feet in height, but at that point, there was no ledge to be concerned about. Yippee! Have another swig! Long live the Muscatel!

diving from the tree

In the evening, they ended up in "GERMAN-TOWN", south

Clearing the ledge

of Brennan, Texas, where the Germans had their Saturday Night Dance and so they "whooped" and "hollered" right along with

Beer party

the German Settlers. They did mostly "square-dances", rather than "polkas" and "waltzes", but that was alright. Except for one thing. Just like everywhere else in Texas, it seemed that all girls over 16 were married and picking up girls was out of the question!

The boys had a ball anyway, dancing and singing and at times, during band-breaks, entertaining the locals with some Dutch songs, well appreciated.

Over ham and eggs in the wee hours of the morning, someone suggested that Brennan had the only legal whorehouse in the area. In no time flat, only Pieter and Frits were left behind, sipping coffee and waiting. Pieter was engaged to a girl back home in HAARLEM and Frits comments were; "If it isn't love, I don't want it and I'm certainly not gonna pay for it!"

Ari-Ari appeared and needed $5 to go back for seconds, so he sold Frits his *FIELDBOOTS* and walked back over to the brothel in his socks!

Things to do on the weekends depended a lot on whether they had a "LONG" week-end or a "SHORT" one, starting on Saturday afternoon.

Texas was too big to travel very far on a SHORT week-end, so things were done kinda close to home.

One Saturday afternoon, the picnic on the base had been most pleasant. Three guitars working away most of the time and the kegs lasted through mid-afternoon.

After 6 o'clock chow in the mess hail, in order to save money, and a pool game or two, six of them piled into Pieter's Studebaker with a case of beer and headed out to "nowhere in particular".

As they passed through the town of Bryan, someone hollered; *"Stop Pieter, back up, back up!"…. "Look over there!"… "Over there!"* There, in broad daylight, stood Harm Appelo, dressed to the hilt. Navy blue suit, white shirt, burgundy tie and black shoes. He stood patiently in line at the movie theater, waiting for the movie to let out. He looked drunk as a skunk, staring at the car with unseeing eyes, but the funniest thing about it all was, he stood in the middle of a long line of people under the sign: "COLORED ONLY", surrounded by some sixty black faces.

The "WHITES ONLY" line started on the other side of the theater entrance, some 50 feet away!

Harm could hear somebody calling his name, but he couldn't focus his eyes well enough to make out where the sound came from, so one of the guys got out of the car, crossed the street and got him.

They would have offered him a beer, but they had no opener, so for the moment they just laughed their heads off about Harm's funny predicament.

Harm was out of it alright. He didn't even know why he was dressed up like that on a hot Texas night and he couldn't even tell them where he'd gotten so drunk. He also had no idea why he was in the movie line, or that he was in the movie line!

It was hilarious

He didn't know what was playing either!!!

According to Harm; "Everybody was in line, so I stayed in line too!"

Too much military training, obviously!

Pieter drove around, looking for a quiet spot and settled for a corner lot, surrounded by trees. From the debris on the lot, they built a camp fire and concentrated on the task of opening beer bottles without a bottle opener. That was an interesting challenge!

Two guys would take a bottle each, hook the "crown-caps"

together and pulled till one of them would come off, while the other got sprayed with beer. Harm was soaked in no time flat and took off his beautiful jacket and threw it someplace. Soon, his shirt was soaked, and inasmuch as he couldn't handle the buttons in his inebriated state, he tore the shirt to threads.

Meanwhile, some serious singing was being done by the campfire and everyone was in "excellent" voice that night.

Somewhere during the evening, it was decided that throwing empty beer bottles all over the place was not civilized, so they improved the situation by imitating a "TRAPSHOOT!"

One guy would throw his bottle in the air, while others threw theirs after it, trying to hit it in mid-air.

Their "hit" averages stunk, but if felt like a refined kinda way to discard the empties.

Just as their averages were improving by the light of a bigger and improved campfire, sirens started screaming in their direction and soon two police cars ran onto the sidewalk, spitting out 4 bulky policemen.

"*Hands up over your heads*!" They meant business!

The boys didn't!

"So nice of you to come. Find a seat and we'll see if we can find you a beer!"

"SHUT UP AND RAISE THOSE HANDS!"

"In our country, we are more polite than that!"

"WE DON'T GIVE A SHIT ABOUT YOUR COUNTRY!, UP, UP, NOW"

"PUT YOUR HANDS ON TOP OF THE CARS!"

"That won't be necessary Mr. Policeman, we believe that they're your cars!"

"SHUT UP, I SAID! PUT THOSE HANDS ON THE CARS!" They were getting mad and rough and started pushing.

"After you, Sir!"

"LIKE HELL! UP, UP!"

"Officer, all we were doing was singing a few songs!, Is that so bad?"

"YOU'RE ALL UNDER ARREST FOR DISTURBANCE OF THE PEACE!"

"Officers, if you came to visit us sometimes we would treat you a lot nicer, you can be sure!"

Meanwhile, the cops felt them down, but other than "hiccups", they had nothing on them.

"GET IN THE CARS!"

"After you Sir, after you!"

"NEGATIVE, GET THE HELL IN!"

On the way to the station house, they asked the officers if they had any special requests and if so, the boys would gladly perform them for'em.

The total ride was but 5 minutes and they were put in a huge cage with one more occupant, a dead down drunk who looked with wide eyes and a dumb look on his face.

He had probably never seen a spectacle like that before in his life!

Seven guys, profusely thanking the police for their hospitality and offering to pay the cab fare.

Once left alone, the songs started up again and at the stroke of midnight, they stood at attention, while they sang the American and Dutch National Anthems.

The policeman on duty thought it was all very funny and kept them supplied with ice water. Nice touch!!

By 2:30 am, everyone was sound asleep on the metal bunks, but by 6:30, they were awakened with a bang!

Bryan jail

Captain Roxs, who couldn't be located the night before because he was out, had been roused by the Officer of the Day at the base, informing him that 7 of his "darlings" were in jail!

There he was, fuming mad. He let out a stream of words like *"CHILDISH", "IMMATURE", "EMBARRASSING", "UNDISCIPLINED", "RESTRICTED"* and more difficult words like that and was gone. Some of the boys weren't even sure if they'd seen a ghost or if that was really ROXS.

It did leave them rather subdued, but when a policeman told them, that they would be fined and relieved, they sighed a premature sigh of relief. Premature, because Roxs reappeared and demanded; "Give me all your money!"

The seven of them had just less than $30 on them. Harm had the most, he had $12.

Roxs fumed even more and left again.

The same understanding cop told them, that their fine was $15 each and Rox was now trying to round up $105 on Sunday morning.

At 8:30 he was back. He took Peter in his car and the rest of them went back in an Air Force van.

They didn't joke this time and the only ones having fun were the police officers, shaking their heads; *"WHAT A GANG!" "WHAT FUN!"*

No fun in the van though! They thought they would go straight to the base brig, but much to their relief, they were dropped at Roxs' office, where they joined Roxs and Pieter Groot.

Roxs was still gloriously pissed. He blamed mostly Pieter, being the oldest and highest ranking N.C.O., but he jumped on everybody and threatened severe penalties for a next time.

As the boys tried to explain that Pieter was not all that responsible and the story started to unfold as to how it all happened, Roxs' exterior started to crack a little.

By the time they got to the part: *"AFTER YOU, SIR"* and *"THANK YOU FOR THE RIDE!"* and *"HOW MUCH DO WE OWE YOU?"*, he completely broke up. When they related how they sang the National Anthems at full "Attention" in the jail cell, everybody just rolled around with laughter and Roxs had tears running down his cheeks.

"Restricted for one week-end" settled things and in retrospect was well worth it.

Of course they had to pay back the money that Roxs had borrowed from his neighbors on Sunday morning, but that was only fair.

On another week-end, Frits had an opportunity to visit Houston with "Joop Tichelaar", the Hollander who was in the Base Hospital at the time of Frits' appendectomy.

Joop's roommate was from Houston and would give them a ride.

Friday night was spent at the roommate's parents house, complete with Bar-B-Que and sightseeing.

On Saturday, Joop and Frits took a bus to Galveston and the beaches and that was indeed most enjoyable. This was their first time at the beach since Coney Island and they swam, flirted, jogged, flirted, swam some more and flirted some more. Great beach!

They ran into a couple of girls from Montana, who had driven their Chevy right onto the beach. The Chevy was old, but the girls were young, but they all had something in common, they were all dressed in black. In their black bathing suits, they were no "ESTER WILLIAMS" but they were pleasant enough and they all ended up in the Black Chevy, going for hamburgers and beer and back to the beach.

Midnight swims and beer had an interesting effect on all of them and they ended up acting like boys and girls together.

It had been a long time since Frits had as much as held a girls hand, leave alone a boobie. Thank God, he remembered how!

They spent the night there under the Galveston sky and the next day the girls drove them back to Houston for their ride back to Bryan.

It had been a very relaxing and enjoyable weekend, but later that week, a most irritating itch started to develop.

At first Frits thought it was some sort of rash in his pubic hair, growing into scabs as he scratched.

At the Doctors office, the Doctor took his tweezers, removed something and dropped it on a white sheet of paper.

"JESUS CHRIST! WHAT IS THAT?" Frits nearly jumped!

"WHAT'S THAT, DOC?"

"NOW THAT'S WHAT'S KNOWN AS A CRAB. YOU'VE CAUGHT THE "CRABS!"

The damn thing was alive and had loads of moving legs, like a bug or an ant.

"WHAT DO I DO NOW?" He sounded panicky!

Well, the solution was simple, but brought a lot of laughter and ribbing.

When his buddies found out, how he'd squat over a newspaper, rubbing himself with D.D.T., listening to the crabs drop on the paper, they really let him have it.

"What a mess. And they were such nice girls from Montana, too!"

One Saturday, the Retired Officers Association, Bryan Chapter, had invited all the Dutchmen to a Bar-B-Que and fishing week end, thanks to the ever loving Capt. Roxs.

Many of the Officers had fought in and over Europe and some even dropped bombs on Holland.

Roxs had arranged for all of the people to get together at the ranch of one of the Officers, where brunch and beer was being served. Afterwards, they'd leave for "LAKE HEARN" and their fishing expedition.

The beer was "iced down" in garbage cans and a hot Texas day can do wonders for an ice cold beer.

Introductions were made, stories exchanged; *"WHERE Y'ALL FROM IN HOLLAND?" "OH, I FLEW OVER THERE MANY TIMES! 25,000 FEET AND COLD AS A BITCH!"*

Others had fought in the Southern part of Holland or had fought with "PATTON" in the "Battle of the Bulge!" ~

Some of the former Officers had fought in the Pacific and could relate to the stories about Japanese Concentration camps, where several of the boys had spent their childhood years.

There was plenty to talk about and many "war-stories" were told back and forth over cold beer and "ribs!"

Afterwards, cars were loaded up with supplies as if they were preparing for "D" day and the guys were divided up into the 7 available cars.

Frits and four more of the "singers" ended up in a '48 Olds Coupe and their host "CHUCK", proved to be the loudest singer of 'em all. Not the best, but the loudest.

"Chuck" was rather chubby and bubbly and had been a maintenance Officer on a B-17 base in England for two years and drank one beer after the other while driving in the convoy.

He had the damnedest habit of looking over his shoulder into the backseat to check on the boys participation in his singing or their reaction to his filthy "RUGBY" songs.

Close to "Lake Hearn", the habit resulted in him running off the road and getting stuck in a ditch.

Thank God, he was not the "tail" car, because it took two other carloads of people to add their muscle and get him back on the road. It was one of the highlights of the trip!

At Lake Hearn, a surprise awaited them. They had never before seen a lake full of dead trees. It gave everything an eerie kinda look. Apparently, it was a man made lake, formed by damming a valley and stocking it with fish after a diverted creek filled it up!

It was different alright! A bare cabin, plain and simple, provided plenty of shelter for the night for everybody and had a long screened porch and a long fishing dock.

A "johnboat" with a 15 HP kicker captured Karel's attention immediately and it didn't take him 15 seconds to start it and run into a dead tree. Ari-Ari, who was his enthusiastic passenger, was pitched overboard in the crash, but wasn't drunk enough to drown.

That heralded in a most pleasant evening.

A big fire got going fast, so a pig could be roasted in good style, although Pieter insisted; "That pig will never lend a proper touch of class to the dinner table, because nobody has brought an apple, to stuff in it's mouth"

Many serious attempts were made at fishing, but with very meager results.

Fewer attempts were made at swimming, but with better results.

It sobered them considerably!

The lake was huge. At one point, they cruised about 4 miles in the johnboat, getting out of the forest of dead trees and into the open water and still they weren't even close to the other side yet. Navigation on that lake nearly proved hazardous to their health, because all those damn trees looked alike and it was nearly impossible to tell where they'd come from. Of course Air Force Navigation Courses did have their advantages and everyone was back somehow before dark.

Singing by the campfire with the help of 4 guitars went over well and some of the "older" Officers showed the "young" studs, that they had not forgotten their drinking days either and could hang in there with the very best!

The ones that had been stationed in England, were particularly entertaining with their "RUGBY" songs, very typically British and very typically dirty.

One song in particular; "*How Ashamed I Was*!" called for a lot of participation and was repeated over and over again, till the Dutch boys knew it by heart as well.

And When I Put It In,,,, Everybody!.... *How Ashamed I Was!"*

"She Said;"You're Mighty Thin"...... How Ashamed I Was!..."

With well-oiled beer-voices, it came off hilariously well! They sang through most of the night.

The pig was a smashing success, the party was a smashing success and the participants ended up pretty well smashed themselves!

The sleeping arrangements were not the greatest, but the boys had known worse, so no complaints were heard, just snores.

When the sun came up, a swim and coffee had everybody up and around again in a good mood. For some reason though, the return trip was more subdued, but that might have been because there was no more beer and most of the boys were not into hard liquor.

Back at the ranch, where they split, promises for life long friendships were renewed and the boys headed back for the base and the books. It had been grand.

One Saturday night, the "FOREIGN" cadets had to provide the entertainment for the show in the Cadet-Club and the Dutchmen decided on an old fashioned "KLOMPENDANS." a wooden-shoe folkdance.

The tune was written out and rehearsed, a harmonica was added for accompaniment and the two dancers rehearsed as often as possible.

The "WOODEN SHOE DANCE" is part of Holland's folklore in the rural Eastern part of the country. ROXS had obtained authentic outfits, complete with wooden shoes.

All of this would have made it "properly" entertaining.

Great swan dive

What made it hilarious was that Karel was dressed up as the woman and Frits as the man.

Karel's big wrestlers frame towered over his "MAN" and the white lace cap on that big crewcut head of his made him look ridiculous. He (sorry, SHE) had huge boobs on that big barrel chest and all of that made him look twice as big as his mate Frits, resembling a skinny little farmer on wooden shoes.

The performance started routinely with the happy pair dancing arm in arm, turning, locking arms again while the CHORUS was singing;

"Als Mien Vader En Mien Moder Noa De Markt Toe Goat, Oh Jaw, Ja So!!!, Meaning; *"When My Father And My Mother Go To Market!.*

They kept dancing and turning and at the point that the choir sang: *"OH JA??, JA SO!!!"*, the charming couple bowed towards one another and then bowed away from each other. Things were going well. The audience was enchanted and clapped along with the rhythm of the dance.

At the last *"OH JA????, JA SO!!*, the whole thing should have ended with an appropriate round of applause, but Karel had different ideas. At: *"OH JA??"*, "SHE" dutifully bowed toward the male partner, but at *"JA SO!!"*, as Frits bowed away from

Karel ten Hove and Frits – wooden shoe dancers

his woman, Karel gave him a shot with his butt and send him flying across the floor on his face.

To the hysterics of the crowd, Karel walked over, picked him up, kissed him on the head and carried "her" man off the floor. What a gas!!!!. Life was a blast!!!

Another time, Austin had to be explored. A little culture would do them good. Pieter had rented a cabin at Lake Buchanan. It was a 4 person cabin, so it would probably hold eight.

Twelve of them went.

Two cars were loaded up on Saturday afternoon and they raced their way to Austin. The road was lightly traveled, which gave them ample opportunity for "FORMATION" driving at 75 MPH.

At one point a "TRANSFER" was necessitated, because ARI-ARI was accused of farting in a crowded car.

At breakneck speed, driving close together, ARI's body was handed out of the Studebaker window and shoved into the Chevy, while half of the guys cheered wildly and the other half watched with baited breath and racing hearts,,, AND..... at that point, they were still sober!!!

Once in Austin, the first and long stop was in "BADEN SPRINGS". What a place!!! A giant pool, crowded with people of all sorts and sizes. All enjoying the hot Texas sun and the cool spring waters. Every guy hoped to make some contact with the opposite sex and to some extent that worked. Girls were aplenty and of course, the noisy group of young fellows with the funny accents and the crazy antics on the diving boards caught a lot of attention.

Karel would make a hand stand at the very edge of the high board, Frits would walk up behind him and while clasping each others ankles, they'd dive in together. Another performance, Frits sitting on Karel's shoulders and diving in together, was stopped by the lifeguard after just one show.

They sure as hell managed to attract a lot of attention though. Karel had a funny stunt. He would go through all the painstaking preparations for a phenomenal dive and after concentrating and deliberately stepping forward as if he would bounce way high for a splendid dive, he'd miss the board, hit it with his butt instead and fall forward in the water. The people loved it.

Invariably, these things ended in *"WHERE Y'ALL FROM?"* and before long, the guys had connections.

Frits was talking with 3 female "friends", charming girls

from Austin, fascinated with the young foreign cadets.

The crazy thing about the threesome was; two of them were married and they seemed more interested in making out than the single one.

Unfortunately, there was no way to pursue that romance that night, because the girls were in one car and had to go. Besides that, the married ones lived at home, another one with Mother-in law. Both their husbands were in the service, one in the Marines in San Diego and the other one in Korea, knocking off Chinese.

The best they could accomplish under the circumstances exchange phone numbers and addresses and Karin, a 110 pound blond in fine shape, assured Frits, she'd write next week. For sure!

The end result was, that all twelve Dutchmen piled back into their cars, without dates, for the ride to Lake Buchanan. They wanted to get there before sundown, but they would have liked it a whole lot better if they had a string of girls along or following them. Tough!

Hamburgers, beer and howling at the moon over the lake created a good atmosphere for an evening of fun.

Ari's *"Don't The Moon Look Lonesome Shinin'On Through The Trees?"* even elicited advice from neighboring cabins. Some of the happy inhabitants meandered over and joined in; *"She Got A Shape Like A Fish, A Face Like A Frog, But I Love That Woman Like A Hollohoohog! Don't That Moon Look Lonesome Shinin' On Through The Trees?"*

The group kept growing and *"Oh, Give Me Land, Lots Of Land Under Starry Skies Above"* was followed by *"Home, Home On The Range!"* and *" Hey, Goodlooking! What's You Got Cooking?"*,all very appropriate, somewhere in the heart of Texas. Good ol' country fun, with lot's a-singin' and a-clappin'!

When some of the goodlooking humans of the other species suggested that there was a good jazz club in Austin, some of the guys were easily dissuaded and piled into the cars with them, heading for the "big town".

It was indeed a treat. A two-story club featured an excellent dance band, rather than a jazz band, but that was even better and the place crawled with people out for a night of fun and laughter.

The drummer was one of the best Frits had ever heard. He could beat out a latin rhytm with his left hand as well as his right hand with such a ferocious pace that it sounded like two drummers were playing at the same time. Just fantastic!

He ate it up and could have stayed all night, but the drinks were expensive and their driver was leaving, so: "back to the lake!"

Disaster struck! Or so it seemed.

Jan Mante had gotten drunk and melancholy, complete with crying fits over his fiancee back home and then suddenly disappeared. The last ones to see him seemed to remember that he went for a swim. Maybe to go and drown himself or maybe just drown accidentally in a drunken stupor.

Peter immediately got in the progress of borrowing a boat and forming a search party.

Earlier in the evening it had been mentioned that on the island in the lake, everyone danced in the nude by a campfire. They wondered if Jan might have wanted to check that out. With no more fire in sight, Frits asked for direction to the island and was told "At ten o'clock to the moon." (Navigational term.)

He was in his bathing suit in no time and hit the water at breakneck speed, stopping every so often to call Jan's name and to listen. At one point, he saw something bopping up and down on the perfect mirror he was swimming in. It turned out to be an empty beer bottle, so he filled it with water and sank it.

At the island, (he hoped it was an island) there was not a sound or a soul. He was about to venture further into the woods, when he thought his name was being called.

He called out "Jan, Jan!", assuming it was his friend calling for him, but it wasn't. It came from the lake. A boat on the lake, to be precise. After a little more calling back and forth, he and the boat met up at the shore of the little island and he found out that now HE was the prime target for the search party. He'd been gone for more than a half hour and Jan had already been found in the backseat of his car.

A belly laugh or so later, the boat was ready to go back to the mainland, but the weather, the lake and the moonlight were all sooooo perfect that he decided to swim back.

Now the moon was directly behind him. The water must have been 80 degrees or more, much warmer than in Baden

Springs, and the night was so beautiful, that it seemed sacriligious to get into a metal boat with a noisy, smoky, stinky engine on it.

He leisurely floated on his back and found it comforting to know that there were no bad after effects from the appendectomy. The exercise didn't hurt at all. *"Good feeling!"* he thought. *"Next week I'll start jogging again."*

Sucked in ? to a T-33

CHAPTER X
THE FUN OF FLYING

There was something about flying!

It might seem inconceivable to some minds that day after day, for more than a year each flight could bring a new experience. Every day there was new excitement and a new eagerness to excel. There seemed no end to it!

Formation flying became the most fantastic thing he'd ever done and the compliments from his peers on the ground were worth millions to him. Remarks like: *"Boy, you guys looked sharp in echelon on downwind and your "pitch-out": BANG-BANG-BANG! Right on schedule to the second! BANG, BANG, one after the other. Looking good, Baby!"*

As each flight of four came in for landing in close formation, showing off their best efforts, they'd head downwind until they were across from their touch-down point on the ground. The lead plane would then bank briskly to the right and and two seconds later #2 would "break", another two seconds #3, two more seconds #4.

They'd gain additional spacing, (hopefully exact spacing) on baseleg and land one after the other, perfectly strung out! That was the objective.

Rittiger had a good team. On flights when Rittiger would ride in Frits' back seat, he'd ask Rittiger to fly the plane a while,

Pretty sharp formation

so he could take some pictures of the formation. They looked as sharp as the "THUNDERBIRDS " and the "BLUE ANGELS", at least in straight and level flight! He had the pictures to prove it!

The hardest position to fly was "#4", the easiest: "#2". The #2 man only had to make minute corrections with the throttle and the stick. #3 was the victim of those corrections and he had to react to #2's throttle settings and stick control. He did not make "minute" corrections, he made "sizable" corrections. #4 was the victim of both their flaws and found himself shoving the throttle from "stop-cock" to full power and back and his control stick all over the place, just trying to stay in formation!

After a while, he got wise to the problem and concentrated on flying formation with the lead plane, letting #2 and #3 bob up and down between them, staying alert for some sudden, "unexpected" big movements.

At airshows, the public "OOHS" and "AAHS" over the announcement that *"They're flying just inches apart at 400 miles an hour!"*, not realizing that that's a very comfortable speed at a very responsive power setting. The relative speed between the planes is barely ONE mile per hour!

Formation flying required all their concentration, all of the time. They couldn't even blink!

They'd line up the "tip-tank-light" on the plane ahead of them with the helmet of the pilot in that plane and stayed there! Dead level with him. Wingtip three feet behind his, just far enough apart that they wouldn't collide, should the lead plane stop dead in the air.

It was the most fascinating job in the world. Hanging in tight with the aircraft next to him (or ABOVE him, if he flew the "SLOT" position) and staying there, even if the background scenery changed from horizon to all blue sky or solid earth, whether they were level or upside down, just hanging in there! Always maintaining the exact position on his lead man's wing. It was a thrilling challenge like he had never experienced before.

Like all other phases of flying or like a new concept in math, all of a sudden, you grasp it, you see it, you feel it and it's a part of you. It's never a problem any more. Just something to keep improving on.

After originally working his fanny off on each flight, all of a sudden everything fell into place very easily.

It happened on their first formation night flight. Just two planes, two students with their instructors. The normally hot and bumpy Texas sky was smooth as silk. That helped. The planes sliced through the air as if they were noiseless gliders and the ride was so smooth, it was like sitting at home in an armchair. Hanging on in tight formation with the navigation lights throwing red and green glows on the silver planes, it became a piece of cake.

Frits relaxed completely, became less tense on the controls and immediately his reactions minimized to the point that the controls barely moved.

Rittiger in back, feeling along on the stick, could feel the change and after a couple of successful steep turns, adding power when the lead turned away from him and retarding the throttle when the lead turned into him, Rittiger encouraged him to "tighten-it-up!" With the wingtip well inside the lead's tiptank and the trailing edge of the tiptank just 6 inches from Frits' leading edge of the wing, his cockpit was only 6 feet from the navigation light on the lead's tiptank. He felt a sense of power he had never felt before. Whether they were standing straight on their nose facing the earth or hanging underneath, facing the stars, he kept that plane right in place, barely moving an inch in any direction. And... ... it didn't seem difficult any more. No… . it was just exhilarating!

He felt regret when they changed position and took over the lead with Rittiger doing the flying. The grand experience ended for the time being.

Once on the ground, he received the greatest of compliments: the other student, who had been in the lead plane when they were meshed close together up there, wouldn't believe that Frits had been at the controls! That was the best type of applause he could have received.

Instrument flying was something different. That was a necessary evil. He had to learn it just like any other pilot had to master it, but it was a royal pain in the neck.

The "link" trainers from their "primary" days had been replaced by "simulators". They were sophisticated electronic wonders. They made flight simulation so realistic, it could make you airsick. Thank God though for simulators, you could "crash", open the door and walk away!

Up in they air… ..one goof and they might bust your butt!

Flying under the "hood", gingerly holding the stick between thumb and forefinger in order to avoid tensing up on the controls, constantly scanning the "flight" and "engine" instruments and keeping them within the range that the instructor expected, was a tiresome chore.

Flying the "RADIO-RANGE" and "homing in" on an airfield could be maddening. On one such flight he got royally screwed up.

Picture a circle, sliced into four equal pieces.

Opposing pieces had the letter "A" in morse code: "DAH-DIT" (— -) or the letter "N", "DIT-DAH" (- —).

When tuned in to the proper frequency, they'd be greeted by "DAH-DIT" or "DIT-DAH" in their earphones.

On their clipboard that was strapped to their knee they could see Radio-Range-Station, the airbase in relation to that station and the "quadrants with "A" or "N". They'd simply fly straight and level for a while untill they reached the area where the "A" and "N" overlapped (————) and thus created an even tone: MMMMMMMMMMMMM.

That was the time to turn onto the heading of that beam as indicated on the chart on their knee. The only thing they didn't know at that point: were they flying to or from the station, because they didn't know which side of the station they were on. However, if the volume of the MMMMMMMMMMMMMM increased, they were heading to the station, if the volume decreased, they were flying away from it. Simple enough. If they were flying away from it, they'd make a 180 and get back on track. Once over the station, all sounds would cease over the "cone of silence" and at that point they'd turn on the heading for the airport and home in on it's beacon. Nuttin to it! Sure!

One hot bumpy afternoon, Frits found the "DAH-DIT", flew until he intersected the MMMMMMMMMM and turned on course. He was very satisfied with himself. The sound did not increase much... .just ever so slightly he thought. It took the longest damn time to reach the station and Rittiger asked on the intercom if he were very far out? He could only answer that they must have been a hell of a long way out, because he still hadn't reached the "cone of silence".

"OPEN THE GODDAMN HOOD!" Rittiger shouted. *"I'VE GOT IT!"*

He laid the plane in a steep bank and asked: *"Do you know where the fuck we are?"*

"DAMN!" There was a training base smack under them, T-28's and T-33's all over the ramp. *"It's not Bryan"* is all he could think of.

"Fucking A it ain't! So what is it?"

"Damn... it looks like WACO based on the runway directions!"

"You're fucking right it's WACO. Now how in the fucking hell did we get here?"

Frits was flabbergasted.

"Check your Goddamn switches!"

Sure as hell, the switch was on "automatic", keeping the sound even, in spite of distance.

"Oh, shit!"

"Goddamn right: oh shit! You could've run out of fuel and bust your fucking ass. Turn the Goddamn bitch around, crawl under the hood and take me home!"

He's muttering to himself: *"Boy oh boy, Why do I have to screw up so often? It seems so stupid!"*

Back on the beam, this time toward Bryan, MMMMMM-MMMM till he intercepted the "cone" and turned on the base beacon frequency. He requested an instrument letdown and a G.C.A. (Ground Controlled Approach) and brought her in over the runway when Rittiger said:*I got it!"* and landed.

He did not get a pink slip, not even a tongue lashing!!! He couldn't believe it.

Maybe everybody was expected to make that mistake once???? Who knew??? HE'd never make that maistake again, that was for sure! He was terribly relieved!

One weekend, Rittiger wanted to visit friends in Nashville, Tennessee and Frits had the honor of flying him there.

Very, very dubious honor!!

Low level instruments, all the way!!!!

2000' navigation under the hood, bumping up and down for six hours. First leg; Bryan-Oklahoma City, then Nashville and back to Bryan on Sunday!

What a chore. First charting the routes with all the headings, speeds, times and frequencies along the way. Then all sorts of listings of commercial radio station along the route, so they would be able to check their progress and course.

At that altitude and in that heat, it was like driving on a cobblestone road with flat tires. It was teeth shattering bumpy! The instruments bounced all over the place, but by having the plane trimmed properly, the altitude and heading were easily maintained.

Changing radio frequencies constantly and checking bearings from two or three stations in order to pinpoint their position at all times, kept him as busy as a one-armed paperhanger!

The important thing was, they got to Oky City without a hitch.

When Rittiger said; *"I GOT IT"*, and Frits could open the hood, the airport was right in front of them. He sighed a sigh of relief. What a job!!

Unfortunately, after refueling, pit-stop and coffee, Rittiger had him right back under the hood again and requested an "instrument take-off!" What a job!

He did more radio work on that weekend, than he had done in his entire year of flying!

Communicating with "GROUND CONTROL" and the "TOWER" on a commercial airport was a "first" for him. Talking with "DEPARTURE CONTROL", "APPROACH CONTROL", getting clearances for every heading and altitude change kept him hopping. Besides that, he had to keep the plane flying!

The same thing happened again going into and out of "Nashville International", but in spite of all the work, it had an exotic touch to it!

It was different. Unmilitary. It was like flying to foreign countries! Invigorating.

In Nashville, they were picked up by a charming young lady and she gave them a brief tour of the city's highlights and landmarks, before stopping for steak, salad and beer.

Frits was dropped off later at a neat little motel, right where the Nashville "action" was. Rittiger gave him $15 beer money and left, which suited Frits just fine.

Nashville was a fine open city with many fine "joints" and even though the music was mostly "*COUNTRY*", Frits enjoyed himself thoroughly! He couldn't latch on to someone to take to bed, but all in all it was a pleasant evening.

By eight in the morning, he was out snapping pictures of the Grand Ol' Opry and the Capitol Building, plus he bought a made to order newspaper, headlining: "FRITS TAKES NASHVILLE BY STORM!", to send home to Mama.

By eleven, Rittiger and company picked him up for the return flight and by mid Sunday afternoon, they parked their T-33 on a quiet ramp of a quiet Bryan A.F.B..

"*Good Flight*!" was all Rittiger had to say and Frits wondered if he was so pre-occupied because he had to face his wife again.

The big "finals" before earning their wings were the "WRITTEN" and "FLIGHT" instrument tests.

The "WRITTEN" was a monstrous test, it was said, so everyone studied night after night and weekends. Everyone but Frits. He was confident he knew it all, so he played pool, drank beer and ran off the base with Pieter in his Studebaker.

When testing time came, he confidently sat down, but coiled back as if he were snakebit.

'Hell, that's not fair!" he screamed to himself. The questions pertained to instruments used in the T-6, the T-28 and other older type aircraft. He had comfortably forgotten about that stuff.

He could recite every word of "*Ol' Man River*", but couldn't remember what made instruments work in a plane he flew just 8 months ago.

He did a lot of guessing and even when the test moved into the "Jet-Age", he was still dumbfounded a lot and consequently, he failed the test.

"Oh boy!", that shook him up. He expected all sorts of repercussions. His ego was totally deflated and he thought that it might mean the end of his flying carreer.

For several days, he was quite subdued. When nothing happened and all he had to do was "reschedule the test", he procrastinated, He went back to playing pool and drinking beer and didn't even crack a book!

It caught up with him again though!

The flying hours mounted up steadily and sooner than inspected, he was scheduled for the all important "instrument flight test!"

Pass that test..... you're a pilot! You've earned your wings!

All the rest was just a matter of ceremony!

Fail it twice...... Goodbye Charley!, Wash-out time!

The flight-test-pilot was a youngish, soft spoken Captain, who spent a considerable amount of time on the ground, preparing for the flight by going over charts and flight plans.

With pencil and paper, figures came out of Frits' brain very rapidly. His calculations and self assuredness must have been convincing to the captain, because he offered no corrections or suggestions. Every so often he'd ask; "How'd you arrive at that?", but that was it! He seemed satisfied!

Instrument take-off at Bryan. The heat was already making it bumpy at 10 AM....Climb out.....intercept Radio Range....fly outbound to Houston...Pick up "Waco" outbound....(good).....

360 degree steep turns,... left... right.... 360 degree standard turns.... maintain altitude..... climbing turns...., pick up Austin outbound.... request let-down and G.C.A. at Fort Hood Army Base.

Everything went well enough, Frits thought, no screw ups at least!

Not so far at least. *"IF I JUST DON'T DO ANYTHING STUPID AGAIN, I'LL BE ALRIGHT!"*

"CLEARED FOR LETDOWN TO FORT HOOD, REPORT AT 2,000" "

Good going! At 2,000' , over to G.C.A. frequency and after some turns and letdown to 1,000', the operator started his routine;

"NO MORE COMMUNICATION ON YOUR PART WILL BE REQUIRED. YOU'RE ON FLIGHTPATH AND ON CENTERLINE. START YOUR DESCENT AT 500 FEET PER MINUTE NOW. ETC. ETC. ETC.....

The outer marker light blinked right on schedule, while Frits wrestled with the plane, that bounced up and down like bouncing ball in the mid-summer heat.

The inner marker beeped momentarily and just a few seconds later;

"YOU'RE OVER TOUCH-DOWN. TAKE OVER VISUALLY!

Normally, the instructor would take over at that point. NORMALLY!!!! NOT THIS TIME!!!!

"TAKE IT AROUND!"

Frits didn't have time to question the order. He shoved the throttle full forward... stopped the vertical speed indicator from going down further.... As the thrust kicked in and started to shove the plane forward, he brought the vertical speed to "CLIMB" at 300' per minute.

Switch to Tower frequency; *"PERMISSION TO LEAVE THE PATTERN"*

He "milked-up" his flaps little by little and increased his vertical speed while maintaining 240 MPH.

He pushed the intercom button; *"WHAT HEADING OUT, SIR?"*

"MAKE A CLIMBING 120 AND COME BACK OVER THE RANGE. THEN HEAD FOR BRYAN!"

"AYE, AYE, SIR!"

He started a slow 120 degree turn and after a few corrections and a few clearances from Ft. Hood Tower, he headed for Bryan.

As they started their letdown to Bryan APE, the test pilot said; *"Come on out! I've got it!"*

Frits slipped back the hood and enjoyed the instructor's aerobatics. He rolled and looped the T-33 a few times and then showed off a perfect 8-point-roll.

On final he said; *You land her! I don't get to do much acrobatics anymore. Kinda miss it!"*

Once out of the aircraft, he didn't say much. No compliments, no criticism, nothing. He just climbed out and walked away, while Frits was going through the final shutdown.

Rittiger climbed on the wing as Frits was unbuckling.

"How'd you do?"

" I dont know, he had me do an instrument go-around atFort Hood and we've never done that before!"

"A fucking go-around on instruments? What goddamn altitude were you at when he told you to take it the fuck around?"

"We were over touchdown. G.C.A.. said; take over visually and he said; take it around!"

"Well, what the fuck did you do?"

"I took it around of course!"

Rittiger was speechless; *"He never done that!"*

Silence for a moment. *"Did he say anything? Good or bad?"*

"No, when we were back over Bryan beacon, he took over climbed up some and did some aerobatics and my, did he ever do a fine...."

"He let you out of the fucking hood?"

"Sure did! And he..."

"You must have done alright! Let me buy you some fucking coffee inside, while we wait!"

They were in the middle of reliving an "instrument go-around" a second time, when word came; *"FORRER, CAPTAIN HICKS WANTS TO SEE YOU!"*

Hicks was Flight "E" commander, Frits jumped! Rittiger went with him.

"Congratulations Forrer, you made it!" Hicks was shaking his hand.

Rittiger started in; *"I knew fucking well this boy.."*

Hicks cut him off with a raise of his hand: *"And my boy, uh, mr. Forrer, you got the highest score ever recorded on this base and for all I know, the highest score ever recorded on any base. Ever!"*

"Congratulations and I'm glad it happened here in "E" flight! And congratulations to you too, mr. Rittiger. Fine work indeed!"

"Dismissed!"

After a snappy salute they were out of the door and Rittiger became hysterical! Standing on a chair, he shouted to the whole "E" Flight crew room, that his student, "FUCK-UP

FORRER", had obtained the highest instrument flight test score ever recorded and then he went on to encourage the backslapping and the handshaking that embarrassed Frits as much as it pleased him.

Later that afternoon, while still basking in the glory of his feat and while shooting pool in the cadet club, his balloon got shattered violently.

Since they were finished with all academics, all they did was fly or hang around. They sat in the showers, escaping the heat while telling jokes under the running water, or hung around the cadet club, playing ping-pong or shooting pool.

When the call came; *"FORRER, YOU IN HERE? YOU'RE WANTED ON THE FLIGHTLINE!"*, he dashed over to the flight line, in spite of the heat! Only the 105 temperature kept him from racing. He just knew he'd won a trophy or something!

Wrong! Wrong! Deadwrong!! He couldn't have been more wrong!

"In Hicks' office!" He was told as he walked in. *"Come in!"*

"Cadet Forrer reporting as ordered, sir!"

No *"AT EASE "*, no smiles, nothing but stern faces.

Rittiger was in civvies. He must have gone home and got called back. For once he said nothing, he kinda hung in a chair.

Even Major Troup was there.

Captain Hicks did the talking:

"You dumb ass, you stupid motherfucking ass!"

Frits turned pale and his heart stopped.

"You failed the fucking written test, didn't you?"

This was the first time Frits had heard Hicks cursing. *"Yes, sir!"*

"And you didn't even bother to retake it, did you?"

"No sir, I figured I still have a whole month!"

"The fuck you do! You pass the written test before you go up for your flight test. Didn't you know that stupid?" Hicks got very red in the face. *"Here you go and fly the best goddamn test we've ever had and it's all null and void, because you dont have your fucking written! Get out there and wait!"*

"YES, SIR!" Frits saluted, about face and made for the water cooler, shaking in his boots. Thank God, the afternoon

class that occupied the crew room had not been there in the morning to witness the hysterics and ignored him, hanging around in his khakis.

"FORRER, IN HERE!"

Back into Hicks' office. The Major doing the talking this time.

"Sit down Forrer!"

"Let me ask you something, cause I dont believe you're as stupid as they've been telling me! Can you go to your barracks, study all night and pass that instrument test in the morning?"

"Yes, sir!"

"Not so damn quick!" He's snapping, all of the sudden!" *Do you know what you're saying?"*

"Yes sir!"

"That you can study enough to pass that test at 8 am?"

"Yes, sir!"

"Goddamn, you're sure of yourself! Maybe that's good! This is what we're gonna do. You pass that written tomorrow, we'll date the flighttest tomorrow and everything will be honky-dory! Can we count on you?"

"Yes, sir!"

"You report to lt. Hendrickson at 7:45 at alpha 4, o.k.?"

"He'll score it when you're finished and inform us. You are going to pass this test, aren't you, cadet?"

"Yes, sir!"

"Dismissed!"

As he's making his "about face", he heard Rittiger for the first time; *"You're fucking A, he's gonna pass it!"*

This time he did jog back! He stripped down to his BVD's and complete with a supply of towels for his dripping sweat, he started.

Sitting, laying, walking, crouching, he was reading and studying constantly in all sorts of positions.

At chow time, the guys asked; *"What happened to you, you never got back?"*

"I'm taking the instrument test in the morning. Let me ask you someth...?"

"You never retook that test?"

"No, but I'll pass it in the morning. Let me ask you something...?"

At least the boys proved to be helpful and answered some of the things, he didn't comprehend. He started with the book from "PRIMARY", with all the forgotten instruments.

It was hard to believe that he had forgotten so many procedures and mechanics of the instruments with which he'd flown just 6 months ago.

He finished one book after the other and started on the sample exams in the back of each volume. In boarding school, back in Holland, he had learned to memorize long lists of foreign words by learning through "association". It was as if he could pull up a page in his memory, scan the page as if he were looking at the actual page, find the proper word and then recite it!

That was what he was working on now with instrument data, rather than "irregular foreign words."

Somewhere during the night he fell asleep, but woke again with a start at 4 AM. He did another sample test, looking up the answers when in doubt. He didn't eat, shave or shower and reported to Lt. Hendrickson at 7:40, who instructed him on the test procedures and told him;

"You have 21/2 hours!"

He started quickly, carefully reading the questions.

"Don't misunderstand the questions!" He told himself. *"You won't pick the wrong answer if you fully understand the question. There can only be one right one and that should be obvious if you understand the question!"*

At 9:15 he turned in his paper.

"Don't you want to take the time to recheck your answers? You have plenty of time left?"

"No, sir! Thank you, sir!" He saluted and left.

He had a splitting headache. A rare one. The only headaches he ever encountered were caused by hangovers. He bought aspirins and breakfast in the P.X. and was asleep by 10.

By 1 PM he was summoned to the flight line and since he had slept in his clothes, he was there in no time flat, slightly rumpled looking, sweaty and unshaven, but nobody seemed to give a damn.

The Major was not there this time. Just Hicks and Rittiger in shorts.

"Sign here, sign here, sign here!" That's all Hicks had to say, it seemed. But,,,, when the papers were completed, Hicks ex-

tended his hand and said; "*Well done, Forrer*!"

"*How'd I do?*" Frits was anxious to know!

"*You passed, right? Otherwise you wouldn't be signing papers now, would you?*" Hicks was grinning as if he were teasing.

"Yeah, but what's my score?"

"*You missed one!, You dumb bastard*!" He burst out laughing and Rittiger joined in. "*Come on, you fucking Dutchman, get into some civvies and I'll buy you a fucking beer*!"

Rittiger swung him around by the shoulder and all Frits could do was; kinda wave at Hicks on the way out.

After beer in town, Bar-B-Que and beer at Rittigers and "JACK DANIELS and coke" in some club that Rittiger belonged to, he was delivered home in time for bed check. STONE DRUNK!

The aspirins came in handy again in the morning, but the headache never left him until he had gotten 4 hours of sleep in the afternoon. Thank God, he didn't have to fly that morning. He might have, crashed!

At suppertime, the questions came. How about that "super" test score? Was it true?

All he could say was; "*I didn't see a score! Hicks said I missed one, but I don't even know which one! I didn't think I missed one*"

The looks on their faces were worth a picture! Nobody really had the nerve to say; "*Quit bragging, you s.o.b., you didn't miss any? Nobody is that good!*"

A lot of them might have thought that, but the rumors came from "Operations", not from Frits.

The "instrument-go-around" was re-lived a few more times with guys worrying themselves silly over something they might never be asked to perform. All Frits could do was review what he had done and it seemed logical, that it must have been the right procedure, otherwise the test/pilot would have jumped all over him.

"How close did you actually come to the ground? Did you actually touch down?"

"No, no, no touchdown!"

As a matter of fact, he had no way of knowing if he had

been within inches or feet from the runway. All he knew was: "GO AROUND!" and shoving the throttle forward to stop the descent and change it to a climb. That happened the moment that he felt the engine kick in and the thrust shoving the plane forward. Come to think of it, he wasn't exactly sure at what point he had brought up the gear. He seemed to remember the climb indicator was reading 500' per minute, his altitude 100' and his speed at 190, but he could not vouch for that. The important thing was, everything was in the "green" before he moved the gear lever "up".

"Did you switch radio's or did the instructor do that?"

"No, no, I did that! I told G.C.A. immediately; "going around!" And I must have been halfway down the runway, when I informed the tower. I asked for climb-out clearance and heading and they obliged ever so nicely!"

"You thought about all that shit, while you were fiddling with gear and flap handles, establishing climbs and directions?"

"Yeah, man! I had nothing else to do!"

"Shit, give that man a beer!"

"Thanks, I think I earned one!"

CHAPTER XI
GRADUATION

Each class was broken down into "increments".

Every six weeks a class graduated, but only one third of the class received their wings at that time. The 2nd "increment" received their wings 2 weeks later and the 3rd a total of 4 weeks behind the first.

The guys in the "first" had to wind up their tests and flying hours and have it all behind them at graduation time. The rest had more leeway.

Frits was in the second increment with Jan Mante. He had all his flying hours in and had nothing left to do, but hang around until the day he'd receive his wings.

The actual graduation ceremony was colorful and emotional and the parade, staged for the graduating class was impressive!

The speeches were proper and inciting to the point, that he felt like grabbing a T-33 and flying to Korea to shoot up some Chinks. The most impressive thing though, was a 1953 OLDS 88 convertible, red and white, top down and sleek! A gift for one of the American Cadets from his beaming parents!

"GOOD GOD, TO EVER OWN A CAR LIKE THAT!"

What an impossible dream for a Dutchman, who was just elevated to Sergeant and now received a whopping $82 a month. What a dream!

Graduation day had been fun, but now he had nothing to do but hang around for two weeks, until his wings would be pinned on his chest.

He was down to $12, so he told Pieter in the Liaison office: *"I'm too broke to do anything around here, so I may as well be broke somewhere else! I'm going to Victoria to see my buddies at Foster Air Force Base."*

He packed a small bag, bought a roll of film, which brought his capital to under $10, and was off.

It was Friday, about noon, as he walked off the base and started hitchhiking, direction Houston.

In Houston, he had a beer or so in some "honky-tonk" joints and checked into the "Y" before midnight.

"Crummy town!" he figured! By 6:30 in the morning, he was hiking his way out of town, destination: Foster Air Force Base, Victoria, Texas.

By one in the afternoon, he was at Foster, but couldn't locate any of his buddies. All out on the town somewhere. He did run into one of the Belgian "Officers in training", who was in the class behind him at Kinston, the Class of 53-G. The Belgian, "GUY", only spoke French, a typical "WALLOON", couldn't under stand a word of Flemish, Belgium's other language.

Just seeing one another again was enough of a reason to hit the Officers Club for a beer and to exchange experiences. Guy decided he'd make some calls and see if he could locate some of the Dutch clowns. Frits made himself useful by hanging out at the piano bar, where a crew-cut 2nd Lieutenant was hammering out one popular tune after the other.

Guy's calls got results. A dark haired, good looking Dutchman walked in the place and introduced himself as Peter van de Meulen-Kunee. Absolutely unknown to Frits at that time, Peter turned out to be a classmate of the "gang",,, Shorty, Ike and Kees. He didn't have any idea where the guys were at or how to get a hold of them on a hot Saturday afternoon.

So they drank beer and talked "flying!"

Peter was keenly interested in the final four weeks of training that he was just about to face and that Frits had just completed. Time flew and beer flowed and Frits was grateful for small favors; he was not allowed to pay in the Officers Club, not being an Officer. It might have been a cockamamie story that Peter dreamed up, but being down to $6, Frits wasn't about to argue.

At chow time they split, Frits heading for the Cadet Mess and possibly his buddies as Peter left to go change clothes. Peter did leave an open invitation to come back to the club later and enjoy their Saturday Night "Patio-Dance".

"SHORTY" and "IKE" were at dinner, because they were restricted to the base. They whooped and hollered and carried on like crazy about seeing their old pal again.

After chow, Frits walked them back to their quarters, where he caught up on all the things that happened since Kinston. Unfortunately, the O.D. broke it up by ejecting Frits from their barracks: " *NO VISITORS, THESE GUYS ARE RESTRICTED!"*

Dumb stiff! Well, the 0-Club seemed like a nice alternative, so in Guy's room in the B.O.Q.s, he showered and changed and was off to the ball.

The band played at poolside and the patio was the dance f1oor. The Texas sky was clear, the air was warm and the beer was cold. So the setting was perfect.

There were very few girls by themselves. All females were escorted, married or single, they all had a man with them. Frits figured it would become more of a "singing" night, than a "swinging", night, but either way, that would do!

At the bar, he was introduced to some of the instructors and once they heard he just graduated from "BRYAN", all sorts of questions popped up!

"DO YOU KNOW, LIEUTENANT SO-AND-SO?" or "CAPTAIN SUCH-AND SUCH?", pilots they had trained with or flown with in Korea. Everyone assumed he was a Pilot and an officer and treated him as an absolute equal. He didn't bother to tell them, that his wings were not due for another two weeks and that his rank would then be Sergeant. It didn't bother him, so why should it bother them? A certain Captain Smith, Peter's instructor, hustled him outside to his table and introduced him all around.

"ANOTHER GODDAMN DUTCHMAN WITH A GERMAN NAME, FRITS!"

One hell of an introduction, but within seconds he was part of the group and was relating stories about BRYAN. One particular point of interest was a rumor that Lt.Col. Jones, Bryan's Ops Officer had hit a tree and Frits was only too happy to relate the story how Jones had clipped some branches on final during a night landing.

All of them seemed to know Jones and all had a good laugh at Jones' expertise, especially about the leaves still being stuck in the wheel well, for the whole base to see the next morning!

That story got him really in with the crowd and when Ann Smith asked him, with a nice southern drawl, if he could dance, he was off to the races! Captain Smith told him:

"YOU KEEP THE OLD BITCH DANCING AND I'LL KEEP SUPPLYING THE BEER!"

Fair Deal!

All went real well until first, some of the guests pitched some others into the pool and second, when he got to be too good at dancing the rumba!

At one point, Ann and Frits were really swaying to "*You Belong To My Heart, Now And Forever*!" and in the wind-up, he dipped her too far and dropped her! She hit the patio stones pretty hard and went out like a light. An embarrassed Frits was trying to pick her up, a waitress was screaming at him, the crowd was "oohing" and "aahing" and Captain Smith hit her in the face with a glass of ice water. Fortunately for all concerned, the ice water did it!

Ann revived with a silly-assed grin on her face and was helped back to her chair. Frits disappeared into the bar, thinking Smith might kill him. Wrong! Within five minutes the Captain was by his side, saying; *"AIN'T YOU GONNA DANCE WITH THE BITCH NO MORE? COME ON OUT BOY, SHE'S GIVING ME ALL KINDS OF GRIEF. COME ON OUT!"*

Rather sheepishly, he made it back out to the patio and found to his relief that everybody thought the whole episode to be rather funny! He had to put up with names like "FRED ASTAIRE", "GENE KELLY!" and "THE DUTCH FLAMENCO DANCER".

At least there was no hostility, just hilarity! He could live with that.

As Smith got drunker, Ann got more romantic on the dance floor and Frits had visions of his head being knocked off before the night was over! It never happened!

By 11:30, a whole gang packed into some automobiles and headed for the "CACTUS CLUB" in Victoria, Texas. It was a private, classy gambling joint with dealers, croupiers and waiters in tuxes and ties.

Ann watched Frits play Black Jack for a while and when his money ran out, she slipped him some more. He was just on a modest winning streak, when he knocked a drink over the green table and Ann hauled him out of the room and into the lounge with the jukebox.

Somewhere during the night, all the beer that he had consumed from 2 PM till 3 AM, caught up with him! He must have conked out. He was sitting in a brightly lit diner at 4 am and he couldn't figure out how he got there, or why? The two Belgian students didn't know how he got there either, but at least they had a car and got him to his cot.

Shorty and Ike weren't allowed to talk to him in the morning, Keesje hadn't come in for the night, so at 10 o'clock, Frits packed his stuff and his hangover and walked out of the gate. Destination San Antonio! "WHY?"

"Never been there!"

He quickly caught a ride into Victoria, but outside of Victoria, on the road to San Antone, he spent two long hours in the increasing heat until a white Chevvy pulled over.

A bleached blond mother of about 35 with a teenage Boy Scout in full uniform picked him up and MAM made the Boy Scout sit in the back, with Frits in front.

A very animated conversation struck up about why he was there, about flying and Holland and through it all the Boy Scout forgot about his early resentment of having to sit in the backseat. He showed a keen interest in everything Frits had to tell and MAM smiled a lot, enjoying the exchange.

She shook him as they were entering the suburbs of San Antonio. He apologized all over the place, when he realized that he had fallen asleep and slid with his head unto her shoulder.

"You must have had a rough night, right?"

"Not really. Too much beer,I guess!"

"Where do you wanna go in San Antone?"

"Oh, I wanna see the Alamo and the river, There's probably a "Y" in the center of town, so somewhere near the center would be fine!"

"You ain't got a place to sleep?" That settled it. They took him home. She had a neat little ranchhouse and while Frits shaved and showered, she fed her Boy Scout son. When Frits reappeared, all fresh and smelling good, she announced she was going to take him out for dinner and show him the town.

She sure took him out alright! To a huge Cowboy Dancehall with music, ribs and pitchers of beer.

She laughed her head off when she heard all about the night before, especially about Ann getting knocked-out on the patio floor. She found it hard to believe though, that he lost her somewhere in the middle of the night and could not remember how or when.

Later on, they walked down by the river, taking in the romantic scenery, watching lit-up boats float by and people strolling hand in hand. They enjoyed some Mexican beer, but had to leave kinda early. She had to get up early to get her son of to school and get herself to work by 8:30, so they strolled back to her car, hand in hand this time.

While in line in a drive-in, waiting to buy a six-pack, they kissed long and hard and his hand slipped inside her bra. Reluctantly, he had to let go of his newfound treasure as the beer line moved up and she had to pull up to the drive-up-window.

He did get back to the booby business on the way back home and had high expectations as to what that might lead to.

Boy, was he in for a surprise!

Back at the house, she showed him the guestroom, turned back the sheets and left, closing the door behind her.

He was thoroughly puzzled! Did she expect him to come to her room? Did she expect anything?

Well, he crawled under the sheets with the light on and was damn near asleep, when she slipped in the door in the sheerest of nightgowns.

Nice little body! Everything there! All in the right places and proportions! She slipped out of her night thing and slid into bed. Some things go well with the lights on!

He didn't remember her leaving, but at 6:30 a knock on the door woke him and she wasn't in bed anymore.

By 7:30, they were on the road and she dropped him off on the road toward the Alamo. They shook hands and she waved goodbye as he thanked her for her hospitality. He couldn't figure out for beans, even years later, whether she just acted cool in front of her son, or because of his performance the night before. The worst part was, he couldn't quite remember!

"I've got to give up drinking so much beer!" he thought. The Alamo was quite a historical sight. He enjoyed learning about it's past and it's famous battle.

As he paid his 50 cents admission, he checked his money.

He had $36.50 in his pocket.

The night before, the lady, (he couldn't remember her name) had treated all night long and he must have been on a winning streak when Ann yanked him from the Black Jack table. He wondered if he had given Ann back her investment? Well,,,, maybe yes,,, maybe no

At least he was fairly solvent at this point!

At one p.m. he was in San Marcos and he wondered if he should try to reach Paolo Brandizzi and Giorgio Morelli at the San Marcos Air Force base. After washing out on Jets, they were transferred to San Marcos for helicopter training.

Standing in the shade under a drugstore overhang, he read 114 degrees on the thermometer outside the store and decided it was too hot to look for anybody! Besides, it being Monday, they were probably in class or on the flight line.

He continued hitchhiking on to Austin, carefully staying in the shade of the trees that lined the road.

He lucked out again. A salesman in a station wagon brought him all the way into the heart of Austin and from a cool Pub, he started calling. He had 3 numbers from that afternoon at the Springs and hit on the first one.

"Karen" was home and *"YEAH!"*, she remembered and *"YEAH!"*, she was anxious to see him, *"YEAH"*, she would meet him, but *"Remember I live with my mother in law?"*. He didn't remember!

"NO!", she would not meet him in a pub! *"People might talk!"*, but she would meet him in a soda fountain at 7.

She gave directions and he went out to locate the place.

After finding the soda fountain, he found a nearby hotel, old but clean. He found out where to get a six pack and returned to catch some ZZZZ's.

6:30 was just right for a shower and a short walk to the

fountain where he was about to order his second milkshake, when she walked in. *WOW!*, she was prettier than he remembered, all blond, Doris Day like! Except more modest boobies.

She didn't want to drink anything, didn't even want to sit down, so he paid his tab and they left.

She wanted to know where he was staying and they went straight for the hotel. No romance or nothing! He had never encountered someone like that!

In the room, they were out of their clothes in no time and sweated up even faster. She had only one hour to spare, so her mother-in-law wouldn't get suspicious. The damn chick wasted most of that hour by talking too much.

They did agree that she would work it out and come to Bryan the next week, when Frits would get his wings and then they would spend a day or two together. Frits would find a place to stay and "Bye," she was gone.

He would have liked to keep her a while longer, it wasn't even dark out yet. He could have worked off some more energy very happily! But she was gone!

Well, the night was young and he wondered if the Jazz Joint would be open on Monday nights, so he hit the streets. It was open alright, but the band wasn't there! Damn! After a few beers and a few more bars, he quit the dull Austin Monday night scene!

By two on Tuesday afternoon, he strolled into ROXS' office and had the start of his life!

"Where the hell have you been? Didn't they arrest you at the gate?"

"You're some dumb fuck-up!" Roxs was learning American!

"Get into your uniform and be back here in five minutes!"

Pieter volunteered; *"I'll drive him!"* and ran after Frits. On the way to the barracks, while he was changing and on the way back, Pieter explained that he was considered A.W.O.L., because he had not completed his flying hours and should have been on the flight line, Friday, Monday and today, Tuesday!

"But I've got all my hours!", he protested to Pieter.

"I'VE GOT ALL MY HOURS!" he repeated it to Hicks, who was getting Rittiger on the phone. *"I've got all my hours!"* he whispered to Roxs, who had driven him down.

Checking the records, he was at least 5 hours short.

"Can I get my logbook? Please?"

That seemed reasonable. Roxs drove him to his quarters and they were back in ten minutes with the logbook. His time checked out. Every hour was accounted for and signed by Rittiger and his monthly summaries were signed by Hicks himself.

An hour by hour comparison of the flight records and his logbook brought forth the discrepancy.... *THE 6 l\ 2 HOURS INSTRUMENT FLIGHT TO O.K.CITY AND NASHVILLE HAD NOT BEEN LOGGED!*

The PILOT-IN-COMMAND Rittiger had forgotten to register the hours with the dispatcher on the Monday morning after the flight! When Rittiger came storming into the office, breathing fire, Hicks greeted him with; ***"YOU DUMB FUCKING ASS!"***

Frits laughed so loud, he thought someone would slap him!

A week later, dressed in khakis and ties, they were "WINGED" in a simple ceremony. Mrs. Roxs pinned on their Dutch as well as their U.S.A.F. Pilots Wings and they all dutifully posed for pictures. Jan Mante, Mrs. Roxs and Frits, beaming from ear to ear. FINALLY!!! After 17 months of training, they were considered legitimate PILOTS!!!

Diplomas in two languages were handed out as well as Sergeant's Chevrons. Frits put them in his pocket.

They received orders to report to WILLIAMS AIR FORCE BASE, TEMPE, ARIZONA. They had two weeks to get there.

Frits figured he'd hitchhike out there and see the country.

Jan had sold his car and was flying out to San Francisco and would make it back to Williams from there.

Karen had backed out of coming and there was nothing left to do, but party locally! Roxs had them over for dinner, beer, songs and jokes. By ten, they hit a local honky-tonk and wailed with the band: *"There Sta-ands The Gla-ass, Fill It Up To The Bri-im, Till My Troubles Are Di-im, There Sta-ands The Gla-ass!"*

By midnight, they stopped for coffee, at "MAJOR'S", a Mexican restaurant in the center of town.

They stopped there frequently and each time, Frits would make a date with MARIA, a pleasant Mexican girl.

She was the youngest of the waitresses there and definitely the cutest one. He had never been able to keep his dates though,

Jan Mante being "winged" by Mrs. Roxs

because he had no wheels and always depended on someone else's transportation. Besides that, by closing time when Maria got off, Frits was usually fast asleep in his cot.

She didn't think it at all funny, when he said; *"Tonight, I'll be back at one. Pieter is gonna lend me his car, right Pieter?"*

Pieter could nod all he wanted. She didn't believe him.

"You always say! You nevar come! " "You lie!"

The boys thought that was very funny, but Maria didn't, she was fighting back tears.

When Frits finished stirring his coffee, he touched her hand with the hot spoon and she blew her lid!

First she tried to hit him with the aluminum ice-water pitcher, but when he fended off the blows successfully, she poured the whole icy content over his head and ran into the kitchen, sobbing.

Everyone who saw it laughed their heads off. Jan and Pieter at the table, others at adjoining tables and even though he was freezing, Frits couldn't help but join in. It was too funny! He was dripping wet and shivered like crazy, but after wringing his shirt in the bathroom and warming it under the hand dryer, things got back to normal.

For a little while at least. When they were checking out a little later, the owner said to two policemen standing near the exit;

"Dat's heem!"

"You're under arrest!"

" Who?? Me??" "What for?"

"Disturbance of the peace! Let's go!"

"Not you, just him!"

Pieter and Jan were ignored and Frits was handcuffed and taken to jail. Same jail, same cell, but a lot less funny!

At 8:30 he was taken before a Judge and fined $13. (2 dollars less than last time!)

He was not released though, had to wait for the "Military transport." The military wasn't through with him yet! *"OH SHIT*!" The military transport, Pieter and the Belgian Major, hauled him back to the base. In Roxs' absence, the Belgian Major took charge and punished him by restricting him to the base for the remainder of his T.D.Y..

Rittiger smuggled him off the base one night for dinner at his house and told him that the Major did a lot of pulling with the U.S.A.F. AUTHORITIES to get him off this lightly. If it had happened one day earlier, before he got his wings, he would have been booted out and his wings would have been denied him. *"You're a lucky shit!"* Rittiger cleaned up his language in front of his wife!

Jan didn't go to Frisco after all and made some short trips and shot a lot of pool. Until one day, late in October, they boarded a plane at College Station for Phoenix, Arizona!

"SO LONG PIETER, SO LONG BRYAN, SO LONG TEXAS!!!!

"IT'S BEEN FUN! ARIZONA, HERE WE COME! LETS HIT IT!"

CHAPTER XII
WILLIAMS AIR FORCE BASE TEMPE, ARIZONA

It was after 7 on Friday evening and the base was virtually deserted. The guard directed them to the O.D.'s office and he knew where to bunk them down.

There was no cadet club and both the Officers club as well as the N.C.O. club yielded no Dutchmen and not much of anything else either. No excitement, nobody with transportation... not much going on!

Back to the O.D.'s office for a bus schedule and a schedule of activities in nearby towns.

They had to wait in line. A lady of about 35 and a young lady of about 23 were ahead of them, talking to a frustrated Officer of the Day. He seemed to be lost for answers and said; *"MAYBE THEY CAN HELP!"*

"THEY" being Jan and Frits, still in uniform from their flight out of Texas.

The dark haired lady extended her hand; *"Hi, I'm Skipper Ross of the Scottsdale Chamber of Commerce and this is Carol and we need young officers, young men I should say, for escorts during the Miss Arizona pageant and need to know who to talk to!"*

"Oh, maybe we can help!"

"I'm Frits, this is Jan, nice to meet you ladies. Let's go where we can talk. You got your car here? Good, let's go!"

" Thanks Lieutenant, have a nice evening!"

In Skipper's car, they headed back to the N.C.O. club, where Jan immediately took the drink orders and Frits seated himself next to Carol and across from Skipper.

"How many men are you going to need and when? And are we going to wear uniforms?"

He was off to the races. He had no idea how many Dutchmen there were on this base, but he was confident that he would have no problem organizing enough guys from ANY nationality for a Beauty Pageant. To make sure that Jan wouldn't say anything to upset the apple cart, Frits told him in Dutch when he arrived with the drinks;

"HOU JE SMOEL!" (shut-up).

And to the ladies; "*Sorry. Bad habit, talking Dutch. I told him I would get the next round.*!" The conversation went from Pageants to "things to do" and dancing and when they discovered nobody had plans for the week-end, the boys changed into their civvies in a hurry and off they went to Phoenix. Nightclubbing in Phoenix was nearly unbelievable at first. It was like coming back home. Nice clubs, liquor licenses, no honky-tonk dumps, well dressed people that took off their hats indoors,,,, it was like Holland. Nice bands, good dance music...They were in heaven!!!

It seemed like an eternity since he had danced. The "patio rumba" with Ann Smith seemed ages ago. Carol was an excellent dancer. She was a teacher, twenty six years old and if she seemed restless at first, she sure warmed up on the dance floor.

On the way to Scottsdale, they continued their warm up in the backseat and at Carol's place, the heat lead to some mutual explosions. Because of Carol's roommate, who was due pretty soon, Frits couldn't stay the night and 'though it left them both hungering for more, Frits left with Skipper and Jan.

Skipper had a nice home in the suburbs of Scottsdale and her eleven year old daughter woke up as her Mam was fixing some bacon and eggs. She chatted happily along with the grown-ups until somewhere in the early morning everyone fell asleep someplace.

The next day, Skipper showed off Scottsdale. It was as pretty a Western town as you could find in the movies and later in the day the boys got introduced all around to the members of the "PRESS CLUB" at the "WESTWARD HO" hotel in Phoenix. That lead to an invitation for "ribs" and "beer" on a ranch behind Camelback mountain and a promise to come back for some horseback riding in the desert.

On Sunday Skipper brought them back to Williams with the understanding, she would receive a call from them as soon as their schedule was determined.

" MANY THANKS FOR A GREAT WEEK-END, GREAT LADY!!!!"

Of all the surprises, there was a Cadet club after all and even some basic training. A Dutchman; "moeders mooiste", Jerry Deerenberg was the Wing Commander and he was rightfully proud of his position, but "oh, so military".

Other surprises; Frits Flieger and Henk Koppelman, whom they hadn't seen since the boat, came hobbling in also. It looked like this was going to be a fun stay!

This was the first time in his $2^1/_2$ year military career, that Frits was transferred without some typical "military chicken-shit" attached. At "WILLIAMS" they got up for "work", headed for breakfast and then on to class or the flight line as if they were civilians. No inspections, lining up or marching off to somewhere. No parades, real laid back. Here they had come to learn how to shoot and bomb. Here they would learn "tactical ground support."

The type of missions flown in Korea day after day in support of the ground troops.

Besides that, "Air to Air" was on the schedule. They prepared to find a "MIG 15", get on his tail and blast him away

It had been three weeks since Frits had last flown in Bryan, but it might as well have been three years, he was so rusty. The instructor might have wondered from time to time, where in the hell he got his wings. He didn't give him a pink slip or scold him or anything, but he did ask; "*Are you sure you're that guy with the high instrument rating?*"

"*God, I was rusty!*" was all Frits could say.

"*You're not kidding*!" That was his only comment.

From there on in, everything went up and up!

Learning how to shoot is the most fun a human being can have in a plane. It combined formation flying with aerobatics and precision shooting. Every flight was a formation flight now. Nobody flew unless it was in a formation of four. The planes would line up on the runway 2 X 2 and as soon as the first element of two was halfway down the runway, the second element would release their brakes and start their take-off run.

Amazingly little was said on the air. Hand signals for take-off, yawing, fishtailing by the flight leaders plane would send them into low-level or high-level battle formations and rocking the wings would have them join back up in a tight formation.

At the gunnery ranges at Gila Bend, close to Yuma and the Mexican border, they would practice their gunnery runs.

First off, everything was done with cameras, no live rounds. The flight would start out in tight formation, then split in elements of two, spread further apart, but other wise the same type of set-up. The leader would have #2 on his left wing, and off to his right would be #3 with #4 off his right wing.

Close to the target, they'd let down and form an "echelon" formation; #1 with #2, #3 and #4 in sequence off his right wing.

The "lead" would contact the range tower and when they were cleared, they'd streak across the range at low level, doing 420 MPH. As they passed over the targets in tight formation, the "lead" would pitch up sharply: "#1 OFF!" and one second later, #2 would pitch-up, calling out: "#2 OFF!". Another second: "3 OFF!" and then "4 OFF!"

By the time #4 called: "4 OFF" and pulled up steeply to climb to 4000 feet above the ground, #1 was ready for his first run. " #1 IN!" He came diving down in a sharp left hand turn toward the target, pulled the trigger as he reached 400 MPH and simultaneously had his target in his sight. Just a few seconds later "1 OFF!", "2 IN"...."2 OFF!"...."3 IN"....."3 OFF"....."4 IN"....."4 OFF"......"1 IN".....and so on. If everyone flew the same tight pattern, the calls; "IN" and "OFF" came with such regular intervals, it was like clockwork.

It took some practice...true...but that was what the training was for. When pulling "OFF", the pilot was nearly on the deck doing 450 MPH, he'd pull a 4 "G" turn, up and to the left and in just a few seconds, he'd roll out at 4000' facing in the opposite direction, about a half a mile from the target. He'd fly straight and level for a while, listening and looking for the other pilots in the pattern. His speed would have dropped to 230 MPH by the time he reached 4000 and he'd let that increase to 300 on his "downwind" leg.

Just a quick crosscheck of all the instruments and switches and it was time to make a 90 degree left turn onto "baseleg" and if everything was perfect, the man in front of him should just be calling: " #— IN". A few seconds later, as he heard "#—OFF", he rolled into his gunnery pass, calling out #—IN". Now he'd concentrate on lining up his sight with the target and ,theoretically, just as his speed hit 400, his sight should be perfectly lined up and all he had to do was squeeze the trigger.

The cameras would roll and later on during the training program, the guns would fire their staccato music for a few seconds and it would be time to release the trigger, pull up and call out: "#... OFF!" ... or..dive into the ground!

The whole procedure would be repeated over and over again, 10 to 12 times. Thereafter, the planes joined up again in formation and headed home. On base, the films would be developed and analyzed. The "HITS" were registered and debriefing would indicate the mistakes that had been made and corrections would be discussed for the next flight.

The sights that were used were not the old "IRON SIGHTS", fixed sights that were used in past wars, their sights consisted of a circle of six diamonds that were "electronically" projected on a plastic sight in front of them.

In the center of the ring of diamonds was a round "PIP", a "greenish yellow" round dot, known as a "TWO MIL PIP". That meant that at a thousand feet, that "PIP" matched a two foot circle. The centers of the huge targets on the ground were exactly two feet in diameter, so when the pilot first rolled in from 4000 feet, his "PIP" seemed immensly bigger than the whole target and the target was the size of a truck.

As he dove in closer, the target seemed to grow and at exactly 1000 feet out, his "PIP" and the centercircle of the target would be a perfect match. Now, if his speed was exactly 400 and his "PIP" was right on the circle,all his bullets would strike the bullseye.

If he was still correcting, crabbing or yawing, his bullets might be all over the place. There were a lot of "ifs."

Winddrift, ruddercontrol, speed, distance out, all of those could affect the "time on target" and the precision with which the young pilot scored.

With practice, they got better at it all the time.

Once they started with "live" ammunition, all that would become more important, because the guns were syncronized at 1000 feet. If all six guns were fired at point blank range, six

Ready to pull some "G's" in a T-33

holes would appear in two rows about two feet apart and each hole 6 inches above the other. So by adjusting the direction in which they fired (syncronized), at 1000 feet all six guns would fire in the same hole. Very nifty!

Too far out or too close in, the holes would not syncronize and the bullets would be more spread out.

So, having the "PIP" right on the target at exactly 1000 feet out and at precisely 400 milies per hour was the ultimate goal. Theoretically, if the trigger was then squeezed for just a few seconds, all the rounds would then be in the bullseye.

Like everything else in flying, if they had nothing else to worry about, like the man in front of them, the plane behind them or how fast the ground came rushing up to them, they'd score 100% all the time.

Worrying about the man behind and if he was about to shoot some hot lead up your tailpipe or if the man in front was too slow and you'd shoot off his tail could definitely affect a man's score. Thank God, the qualifying score was only 15%.

To make it more interesting, HIGH-LEVEL and LOW-LEVEL gunnery was introduced. Coming in on the target at LOW-LEVEL, they'd just about be scraping the cactus and in HIGH-LEVEL attacks, they'd nearly stand on their heads, screaming down.

Bombing was introduced just about the same way. Huge target rings were painted on the ground, measuring 25', 50', 75', and 100' circles in which they were to drop their bombs from various angles.

It was fascinating fun, but also very exhausting! Flying several missions a day, pulling 4 "G's" constantly on those gunnery and bombing runs could wear the hell out of one's arm and in spite of the G-suit, the body took a pounding.

After quite a number of "movie" missions, the young pilots were ready for LIVE attacks. The JOY-STICK had quite a number of buttons on it and a few more of them would now see service. The stick was designed to fit a normal size hand comfortably. On top was a round button, a nickel sized "cooliehat", which was manipulated by the thumb in order to balance out the aircraft through it's aileron and elevator trimtabs. That helped ease the pressure the pilot had to exert in order to control the plane.

On the bottom of the grip, controlled by the pinkie was the radio button and until "gunnery", those were the only ones they had used to date.

The "trigger" was right where it belonged: under the trigger finger.

The "bomb" or "rocket" switch was on the top, next to the "trim" button and whether it released bombs or rockets depended on the "selector-switch-setting" on the left panel behind the throttle.

When all the switches were activated, the pilot could simultaneously "TALK", "TRIM", "SHOOT" (guns or rockets) or "BOMB".

Because most people do not have two thumbs on the same hand, the smart thing to do was to "trim-out" the plane before starting a bombing run.

On the very first "live" mission, on the very first "live" pass, after turning the gunnery switch for the very first time from "camera" to "fire", he screwed up already. He was #2 that day and after the "pitch-up" and switching to "fire" on the downwind leg, he rolled onto baseleg just as the instructor in front of him called out: "#1 IN!"

Rolling the plane sharply on it's side and pulling it into a tight turn, he heard *"TATATATATATA"* .

He had inadvertently put pressure on the trigger that was now "HOT" and fired off a few rounds.

He immediately released the pressure on the switch and rolled in for his gunnery pass: "#2 IN".

This was to be the first real "LIVE" gunnery pass in his life. He'd seen dozens of them being made by Allied fighters during World War ll, but this was finally his own shooting experience. The Range Tower had different ideas: *"Who fired on downwind?"*

Frits continued his pass, fired his rounds, called: "#2 OFF!" And pulled up to the perch.

"Who the hell fired that round on downwind?" This time it was the instructor calling. For a second or two Frits considered not saying anything, but couldn't. *"It was number two!"*

"Number two, break out of the pattern and return to base!"

"Oh boy!" "Number two, roger, returning to base!"

"OH BOY, O BOY! Screwed up again!" The flight back to base seemed to take an eternity. "Why was he sent back so briskly?" He didn't understand. Just a few point 50's fired into the mountains? What's the big deal? It was stupid! True! But would they kick a guy out of the service for that? Would they?

He worried all the way home and couldn't wait for the rest of the team to land so he could face the music and get it over with.

"MUSIC" was right! But it was not "music" to his ears, that was for sure!

"Were you trying to shoot my ass out of the sky over there?"

"No sir!"

"You fired right behind me, right? Well, I didn't get shot down over Korea and I damn well wont get shot down over Arizona! Pink slip! Failing flight. Give me your logbook."

It was very depressing, but nothing else was said about it and life and flying went on as usual. The flying part was so much fun, that it seemed inconceivable that someone could do this and get paid for it too.

All the drudgery of military life had faded away. The guys virtually lived in their flight suits during the day and their civvies at night. No military chickenshit whatsoever! The normal courtesies: "Yes Sir!" and "No Sir!" would have been extended in civilian life as well.

As time progressed, the little smoke bombs were replaced by 500 pounders and rockets would be either a whole slew of 2.75 inchers or half as many 5 inch rockets. Instead of just 2 loaded guns, they were now flying with 6 loaded machine guns, spitting massive amounts of 50 caliber bullets at the targets.

Dropping bombs held no sensation for the pilot. There was no feeling whatsoever when the bombs released, whether it was a little smoke bomb out of a pod underneath the belly of the aircraft or two 500 pounders from underneath the wings.

It was quite different with rockets and guns. The rockets made very little noise, but to see those continuous puffs of smoke outrace the airplane to the target was sensational and contrary to the bombs, the pilot could actually see them impact.

Bombs didn't hit until the pilot had pulled well off the target and even if he tried to see them by craning his neck, he hardly ever saw the explosion and he had to rely entirely on the Range Tower to report the accuracy of his hits. (or the lack thereof). Rockets hitting in front of him as he's screaming down to earth gave the pilot a real sensation of being involved in warfare.

Six 50 caliber machine guns firing together, was an altogether different experience. The kick-back that a hunter feels against his shoulder when he fires his gun, was multiplied a hundred times when six guns did their *RATATATATATATAT*. It felt like putting the brakes on. The noise and the gunpowder smell brought out real hunter instincts. Going in on those targets at blazing speeds with all 6 guns blasting away, made them feel like real fighter pilots. They couldn't get enough of it.

On the ground, life was beautiful. The amount of studying and ground preparation was minimal, but it was fun anyway. How could anyone possibly resent spending time on their favorite subject: "Shooting and bombing the hell out of things?"

Henk Koppelman had a record player and entertained day and night with DAVE BRUBECK's *"Jazz Goes To College!"* with PAUL DESMOND on sax and CHICO HAMILTON on drums. As far as Frits was concerned that was just about the finest JAZZ that had been produced in years.

The plans for the SCOTTSDALE BEAUTY PAGEANT were coming along well and Skipper introduced them to the MISS SCOTTSDALE contestant, for whom Koppelman went absolutely "GAGA!"

A clique formed again consisting of Mante, Koppelman, Flieger and Forrer and most week-ends they could be found in or around Scottsdale.

It was fun to watch people's faces as they were introduced, especially when people heard; ***"This is Frits"*** and ***"This is Frits!"*** Most people thought they were being had! Let's face it! How often does one meet a Frits? And then two in the same room?

On November 14, Skipper arranged for a birthday party for Frits at the home of the famous poetess PATRICIA BENTON somewhere in the desert. The boys were flattered, but had to admit they had never heard of her before, nor ever read her poetry, but that was soon corrected. She presented Frits with an autographed copy of her newest book, "THE YOUNG CORN RISES", a collection of poems based on old Indian chants. Frits cherished it immensely. This was his third birthday in the service already. His 20th in the brig in Deelen, his 21st in Kinston N.C., after solo-ing the T-6 and now all the way in Arizona!. He was duly thrilled.

Henk bought an old "HUDSON", a '39 with an electric shift. It cost him all of $75 but it gave them immediate mobility and freedom. It was also easier on Skipper, she didn't have to come all the way to base anymore in order to collect them. The darn thing was old, but it was stately and big. They loved it!

Until that date, he had never been to Mexico. Neither had Frits Flieger. With 5 days T.D.Y. coming up, they decided this was their chance and planned to hitch-hike down to "NOGALES". Flieger had all of his flights in, but Forrer had to fly

22nd birthday – Frits, Margie, Mante, Skipper, Koppelman

one more "ALL OUT" mission. Just one more mission with bombs, rockets and all 6 "point 50's" loaded to the max, early in the morning and they'd be off to "MEXICO". They would leave their luggage in Phoenix at the bus depot, hitch-hike to Mexico for a few days and pick up their stuff on their way into "LUKE A.F.B." on the North West side of Phoenix. They were excited about leaving. Everything was packed and ready to go, just one more flight, check in and out and *"VAMANOS!" "VIVA MEXICO!"*

Everything was routine. Flight of four. Frits was # 3 man. The schedule was; " 2 bombing runs, dropping 500 pounders, 4 rocket passes, firing 5" rockets, 4 gunnery passes, firing 6 point 50's and return home. Nuttin' to it!

Early take-off while the air was still cool! Great! Later in the day when it got hot in the desert, a plane loaded like that could hardly get off the ground.

#1 element, (#1and #2)....OFF AND RUNNING!....Halfway down the runway.......#2 element, (#3 and #4), Frits in the lead, moving in unison, shoving the power to 98%.....started their take-off run. Heavy as they were with all that armament, they accelerated beautifully, following the smoke trail of the lead element.

With hand signals, the gears were brought up simultaneously. The flaps followed and they changed from "TOWER" frequency to the "COMBAT" channel. At low altitude, he started a turn, trying to cut off the Flight Leader and thus catch up with him quickly. All went smoothly. "Lead" asked for a "radio-check" on "F" channel and one after the other called in; "2 check", "3 check" and "4 check". Everything was "HONKY-DORY!"

"LOW LEVEL BATTLE!!!!" They spread out. Frits and his wingman fell back into his "low-level" position. No sweat! Over the target right on time.

"TIGHTEN UP!" The planes formed a tight "ECHELON" formation and "zoomed" down on the deck toward their targets. *"ONE IN!", "TWO IN!" " ONE OFF", THREE IN!"TWO OFF!",FOUR IN!" "THREE OFF!'* and *"FOUR OFF!"* Set bomb switch, check instruments ....GUN and ROCKET switches "OFF"....Bomb switch; "ON".

"ONE IN!"....Frits could see him rolling left while "#2" was turning "base" up ahead. The team was good by now, their timing: ...."Fine-tuned!" They were exactly at 4000', #2 was rolling in right on schedule....Frits didn't hear him....,maybe it has gotten to be so much second nature to hear that sequence: " 2 IN!" that it didn't sink in, but he also didn't hear the Range Tower report #1's hits and neither did he hear;"#1 OFF" and "#2 IN"... ... or "#2 OFF!". He could see #2 pulling up off the target ahead of him, but no "#2 OFF!"" sounded in his earphones. Well... . he was clear... . so he rolled into his pass... .. "#3, IN!"... . *"TOWER, THIS IS #3, DO YOU READ?"????*

No answer!!!!

He pushed the "BOMB" button on schedule and pulled up, rocking his wings, meaning "NO RADIO!"

Instead of pulling up the "perch" in a 180 degree turn, he continued to climb straight out, "as per Radio Failure" procedure, the way they had been taught!

Away from and above the range pattern, he checked; *"Radio switches"..."on!" ..."circuit breakers...."in"..."channel..F"...*

"Electrical gauges"....."oh... . shit!!!!"

No generator output... Before take-off.. it was alright...

Now....no generator!!!...No electrical power!!!....other than the batteries....and they must have gotten weak, otherwise the radio wouldn't have given out....

"Electrical emergency!"

#1 rule; Land at "GILA BEND!"

In the middle of the most GODFORSAKEN, snake infested desert in America was a slab of concrete called "GILA BEND AIRPORT." Just a runway, nothing else! A Danish student sat there just last week in the shade of his wing until he passed out. It had taken a T-6 4 hours to get to him in order to rescue him!

"Screw that!" "I'm going to Mexico! First;....drop rockets and bombs!!!! Don't take them home!"

He pulled the "emergency bomb-release".......and...KNEW NOTHING!

There was no way of knowing if the rockets or bombs were still there or NOT!

"Damn!"

"Turn off everything electrical and climb!"

As long as the batteries would feed the fuel pumps, they would keep filling the main-tank behind the cockpit and that one was gravity-fed.

He left the fuel-pump switches on, while climbing towards Phoenix. 25.000' was high enough. Even if he flamed out right then and there, he could glide to a number of airstrips in the area.

The airspeed indicator, altimeter, vertical speed indicator, needle and ball, and the magnetic compass still worked because they were not electrically operated.

He started to relax. Way up there, in the clear blue beautiful morning, he started to calmly review everything.

"Oxygen?... O.K.!... .No sweat!!!...Generator...Check Circuit-breaker once more!!!!....No good!!...All other Electrical Switches and Circuitbreakers "Off"....

With the "trim" power and the "aileron booster" OFF, the plane handled very sluggishly, but he could handle that!

When Phoenix came into view, he reduced power to conserve fuel and started a slow descent toward "Williams". He had to keep his eyes wide open for other aircraft in the area, who might not have any way of knowing about his presence. He assumed that the range-tower had informed "Williams-tower" to expect a plane an inbound plane without a radio, so he was confident, that they were expecting him.

If his bombs and rockets had not dropped off, he would be very heavy coming in and he would have to come in fast and short.

If his fuel indicator showed a "FULL" main-tank when he came close to the base, he would fly a "HUNG-BOMB" pattern, making a big square around the entire base, so any falling armament would not land on any of the base buildings.

If he was "short" in the main-tank, he'd make a straight-in and rely on the tower to declare an emergency and clear all airplanes out of the area and off the runway.

The hydraulic pumps were "engine-driven", not "electrically", so his "landing gear" and his "dive-brakes" shouldn't give him any problems. He checked the circuit breakers... . "gear..IN" ... "dive brakes... . IN". They wouldn't use any juice until he hit the switches.

Best to be prepared!

Over "WILLIAMS" at 6000'.

He saw planes taking off. They had switched runways!! Wind must have changed. *"NO SWEAT!"* Power at 80%, gliding nicely.

"INSTRUMENT CIRCUITBREAKERS "IN"!" The needles were slowly moving back into position!...very slowly... His batteries must be about to give out.... The main tank needle moved up slowly to about half and he's satisfied that he's got aplenty to make it in with.

Circuit-breakers "out" again!

The wing tanks might still be feeding some on those dying batteries, but he didn't need them.

He completed a slow, wide turn, continuously looking for other aircraft and down to 1000' he lined up with the runway.

He inched the power up to 84% and started his run smack over the runway. He rocked his wings from one end of the runway to the other, indicating; "NO RADIO". The tower would do the rest.

Left turn... . another left turn... way beyond the buildings... .. The tower would understand; "HUNG-BOMB-PATTERN!"

He could concentrate on getting the plane down,,,, the tower would clear the field.

"DOWN-WIND-CHECK", "10 degrees of flaps...." *"OH SHIT!"*

The flaps barely moved!!!!! *"RIGHT... . RIGHT... . HY-*

DRAULICLY OPERATED, ELECTRICALLY ACTUATED!!!!" They weren't going to do a damn thing! Switch back up! *"BETTER OFF THINKING "NO FLAPS!"*.

That meant an even faster approach!! That should be interesting!!!

He lowered the gear! He thought he heard them locking into place, but the indicators just "TILTED", they didn't "LOCK".

He shook the plane by jamming his rudder pedals from one extreme to another, hoping something would lock if it hadn't before.

He felt pretty confident because as long as his engine was running, his hydraulic pressure had to be O.K. and his wheels had to be locked, in spite of what his gauges were saying.

Left turn onto base-leg... . maintaining 220 MPH with power... . 800' up... . he was in good shape... . He was kinda far out, but that was the whole idea.... No other planes in sight????

Runway clear????? He turned "final" at 500'...He was sitting pretty... . AS LONG AS HE HAD POWER!!!... . If he "flamed-out" and had to "belly-in" before he reached the runway, he'd crack up the gear, land on top of his own bombs and rockets and he would be in "PILOT-HEAVEN" instantly! ALL HE NEEDED WAS 30 MORE SECONDS OF POWER!!" Coming in at 200 MPH, loaded with bombs and rockets, he was like a flying time-bomb!

The runway came virtually shooting at him and well before he reached the "over-run" he cut the power to 80%... . then 76%... . nose coming up... . power to 80% again... 170 MPH... ONE foot above the ground with his nose high... . past the over-run... over the runway... *"TOUCH-DOWN"... . THE MAINGEAR IS HOLDING!!!!... "BACKSTICK.... BACKSTICK... ALL THE WAY BAC... , HOLDING OFF THE NOSEGEAR... ."*

He "stop-cocked" the throttle, shut off all switches, just in case he ground-looped... . Lessen the chance of fire...

The nose wheel dropped easily as his speed ran out and he started to tap the brakes. *"GENTLY, GENTLY!"*, he talked to himself, *"DON'T BURN OUT THE BRAKES!!!!!"*, he applied the brakes,,, released them,,,, applied them again... keeping the brakes from burning up underneath him. He rolled till the very end of the runway and as his canopy opened slowly, he turned off on a taxi strip and slammed the brakes! In seconds he was out of the cockpit, gathered his chute and charts and was off the wing, dashing away from the plane.

Ambulances and fire trucks were bearing down on him the minute he hit the ground and a paramedic was checking him over within 20 seconds. *"Y'ALRIGHT?"*

" I'm fine, thanks".

To a fireman; *"The plane's alright! Just the generator, that's all!"*

"Good show! Good show!!!!" A major in a staff car was pumping his hand, *"Fine landing, son! You o.k.?*

"Yes sir, thank you sir!"

"Were you worried?" The major asked pointing at the bombs and rockets underneath the plane.

"Concerned? Yes!, Worried? No! I knew she'd bring me home!"

"She brought you home?? Shit, you brought her home! Let's go to de-brief, we wanna hear all about this! Good work son!"

Because the base had been put on "EMERGENCY STATUS", the de-briefing took place in the tower and official reports had to be filed. They questioned him endlessly.

Step by step, they went over the entire flight. From the first "#3 In" until he crawled out of the cockpit and they couldn't find any flaws except ONE! He was complimented extensively about all of his actions until they hit him smack between the eyes..*!!!! "HE FAILED THE FLIGHT AND WOULD RECEIVE A PINK SLIP FOR FAILING TO OBEY A DIRECT ORDER. IN AN EMERGENCY: LAND AT GILA-BEND!"*

He was restricted to the base until further notice!

"HOLY SHIT! WHAT A MESS!"

Other than that he was congratulated on an excellent piece of airmanship. An example for all to see!!!!

"BIG STINKING DEAL!"

He found Frits Flieger and told him to "stand-by!"

His Flight Commander complimented him on an "outstanding" landing and told him to wait till his Instructor returned.

"THANK YOU… . I THINK?"

The instructor jumped all over him! They had flown over Gila-Bend and off course: NO PLANE! He had looked all over the desert for a "fireball", but again; NO FIREBALL! He had fully expected Frits to bail out and drop his plane somewhere in the desert!

An hour dragged by and finally; *"IN THE WINGCOMMANDERS OFFICE!"*

"Because of remarkable airmanship,,, because it was your last flight,,,,last day at Williams,,,,,pink slip,,,,,restricted to base until T.D.Y,,,,,commended as a pilot,,,,,"

12 Noon, Frits and Frits were on their way!!!!!!

CHAPTER XIII
"VIVA MEXICO!!!!"

Hitchhiking with just a toothbrush and some money.

Their timing was great! It was November 25 and all American G.I.'s and Cadets were broke, wouldn't get paid till December 1, so Mexico would be deserted.

A few short rides got them out of town and the next ride got them all the way to Tucson.

When they passed DAVIS-MONTHAN A.F.B., they considered stopping in Tucson for the day. W.W.II planes as far as the eye could see: B-17's, B-29-'s, B-24's, P-47's, P-40's... .. planes and planes and planes. They just had to come back and tour the base! But today...on to Mexico!!! Maybe on the way back?

The heat was bearable, 95 and dry and they made good progress. By 5:30 they walked across the border from Nogales, U.S.A., to Nogales, Mexico.

They had been advised to act like Americans, no passport, no I.D.'s. GRINGO's in trouble were simply returned across the border. FOREIGNERS ended up in jail and stayed there until diplomatic channels had been engaged and that could take months!

The one Frits wasn't worried about those prospects at all, the other Frits was duly concerned! He had experience! He said; *"YOU NEVER KNOW!"*

He should know! He had experience!

First things first. A clean, modest hotel room. They hid most of their money in the room and parked their toothbrushes.

Next a Tequila, complete with salt on their hands and a cold "cerveza" for a chaser. 6 PM was certainly a good time to start any cocktail hour, so it was most appropriate that they'd start theirs. The hotel bar was practically deserted which had a positive side effect. The bartender had all the time in the world for them and he gave them plenty of free advice. Where to go, where NOT to go, where to get the best "PESO" exchange, to *"always pay weez pesos", "ees much cheaper"* and *"girls weell steel less"!* Also advice as to: where to eat and what to eat.

They wanted to eat with the Mexicans, not with the tourists.

He told them all about the raunchy nightclub shows, what to look for and what to look OUT for!

"Girls?" Plenty! No Gringo's in town, so plenty of choice. $2 for a "queeckie", $5 for "all-night!"

"No, no, we're not going to pay for sex! For love or nothing, that's it! No pay!" They shook hands on that one! *"NO SEX FOR PESOS!"*

One more tequila on successfully completing *"GUNNERY, PHASE ONE!"* One more tequila for the impending success of *"GUNNERY, PHASE TWO!"* coming up and one more tequila to the health of the great bartender and they were off to eat.

Frits Flieger was born in Indonesia and loved all the hot spices and sauces that came with the Mexican meal and he used an awful lot of it. It was amazing he didn't burn his guts out, although the cold cerveza probably kept the flames down some.

They ate till they were stuffed, but that offset some of that tequila and solid proof of that was that Fliegers tongue seemed less thick at the end of the meal than it had been when they walked in. The cost of the food and beer was near to nothing by American standards, so there was no reason to hold back. And they didn't!

While there was still some daylight, they wanted to tour the city and see the sights but that only took ten minutes.

Before they had gone five blocks, they were sitting at a bar in a nightclub with a cool bottle of XX beer in front of them and 15 girls surrounding them. There were no other patrons in the joint. The girls begged for drinks, pressured to come with them for just $2, even $1.50 and their hands stroked their thighs while pressing their boobies into them from all sides.

"This is what it must be like in a harem!" Frits ONE told Frits TWO in Dutch, *"but let's get outta here and take in a show someplace!"*

"No fair what you speak!" The girls didn't like the idea of them playing the same trick that they pulled themselves all the time; speaking in a tongue that the fellas couldn't understand.

"But I like it here! And I think that I'm in love!"

"Come on, you shithead! You'll be in love ten more times before midnight! Let's go before they strip you clean!"

Reluctantly Flieger pinched a few more boobies and joined him on the way out.

"I like it in there!" Flieger is still complaining!

"I know you do, but I also know you'll like the next ten places we hit and in all of them they'll give you all that attention until you're broke or a "Yanquee" comes in with more money than you!"

"O.K.,O.K!, Where's the show?'

The show didn't start till four beers later and the scenery hadn't changed much, just the names of the clubs. The shows were dirty, but not as filthy as had been predicted by some of the "experienced" guys on the base. There was no "PONY" or "DOG" sex, but there might as well have been because the barflies had their hands right in the guys' crotches and as the hour got later and business didn't pick up, the girls got fresher by the minute.

Bare boobies were stuck right in their faces and nipples were offered to suck on right at the bar or in a booth.

Flieger was in love again! He came over to Frits at the bar with his "BEATRICE", or "BAY-AH-TREES" as she pronounced it. He danced intimately with her, bought her drinks, but resisted going to her room for the $3 that she wanted.

The girls in the place outnumbered the men 3 to 1, which was typical for that time of the month until after the 1st, when the figures would probably be reversed. The girls therefore didn't give up easily, there were no other customers to turn to!

Forrer would buy them a beer after a dance, but remained adamant about paying for some "nookie!"

Flieger, getting more soused by the minute was in rare form again.

He only had eyes for BEATRICE, even though there must

have been at least twenty girls in the bar that were better looking or at least in better shape. At one point he tearfully joined Forrer at the bar and told him about his unabashed love for his "BEE-AH-TREES" and how he wanted to go home with her and how much cheaper it was to spend $3 in bed than to stay in the bar and keep buying drinks for her. He exclaimed that he had never loved anyone like that before in his whole life and that financially and emotionally it didn't make sense to stay in the bar any longer and he would be better off going home with her.

Sound reasoning!

Being reminded of their pledge only brought more tears and beers and within the half hour, when her price had dropped to $2, he came back to the bar; "*Screw you Forrer, I'm leaving*!" and supported by the love of his life, he left.

Now the club was down to just one "FRITS" and that one decided there had to be more shows in town and more clubs. One thing had become abundantly clear though: "There were no girls in that border town looking for love for love's sake!" The only tourists were male and the women were all hookers.

One cute little thing tried to charm him and turned his stomach instead. She was pretty as a picture, obviously new at the trade because she barely spoke a word of English and still had an aura of innocence about her. When she kept buttering up to him and lowered the price to $1.50, he asked her how old she was. When she said: "15", he got nauseated! All he could think of was his 15 year old sister at home!

The very thought that his little sister Hedwig would have to hustle for her daily bread like this, turned his stomach.

By two AM he was so full of beer and tequila, that he wobbled his way back to the hotel and barely made it to his bed.

Morning, hot dry wind, hangover and thirst all arrived at 9 o'clock. His throat was so dry that he thought he swallowed part of the desert. The bartender advised "cold beer", "spice omelet" and "scalding coffee" as a cure and indeed, they performed wonders. By ten he was touring the town, recording the highlights for posterity.

He couldn't figure out for the world in what bar he had last seen Flieger. The one bar of which he was "sure" that he had seen the love-sick Flieger with Beatrice, denied knowing any girl by that name so he roamed the residential section, taking pictures of the shambles that these people lived in.

Frits Flieger and Beatrice

By one o'clock he was just drowning a tamale in cold beer when he saw him. From his vantage point at the restaurant window, he saw Flieger coming down the street hanging onto two Mexican beauties. The waiter nearly tackled him at the door, thinking his customer was running out on his check. With the waiter still holding on to his arm, he stuck his fingers in his mouth and whistled. Flieger looked up and around, but it took a second whistle for him to spot his pal and with his arms around the girls and a big shit-eating grin on his face, he made for the restaurant.

The 5 of them, waiter included, moved inside and once it dawned on the waiter that his "cuenta" was not endangered, he hustled up chairs and drinks with the friendliest smile in town.

Flieger seemed stone sober, but in rare spirits! These two girls were the "loveliest", the "greatest" and the "sexiest" that he had ever known in all his life. The way he said it, it sounded like he'd been alive 1000 years! The other Frits had to admit that they were most charming and the fact that they couldn't speak but two words of English meant at least that they were not experienced whores as yet.

"DE DONDE ESTA USTED?" was about the extent of Forrer's spanish, but it provoked such a longwinded and rapid fire answer that he could not make heads or tails out of it. With a napkin and a "lapiz" from the waiter he drew a rough sketch of Mexico and the girls pointed out a spot near Mexico City as to their town of origin. If he caught it correctly, they were cousins. Their Mama's were "hermanas". *"MUY BIEN!" "QUANTOS TIEMPE TIENEN USTEDES AQUI EN NOGALES?"*

In Dutch, Flieger interrupts: *"How do you know all that shit?"*

"Talent, boy, sheer talent!"

"DOS SEMANAS?" "Hey Frits, They claim they've only been here for two weeks, where'd you find them?"

"Remember Beatrice?"

"How could I forget?"

"Well, Beatrice and I had a fight. She wanted more money or I had to get out, so I got out. I went for breakfast next door and that's where I met them.

"They live right there with an aunt or something, but anyhow, these girls are no hookers!"

"No, I'm sure they came to this resort town for their health!"

Meanwhile they all ate and drank and the girls seemed

to have all the time in the world. Their story was probably a familiar one: Lots of poverty at home and a chance to move up here and make some money. Frits was hoping they'd tell him something about: factory work, grocery store clerk or something else respectable, but they just smiled; *"NO TRABAJO CON LOS MANOS! NO,NO,NO! TRABAJAMOS CON "LOS GRINGOS!"*

They had no honorable intentions whatsoever!

To them, making money was the only honorable thing to do!

When he explained to Flieger that the girls' intentions were the same as those of any other chick in Nogales, Flieger defied him and challenged him to ask; '*How much for an hour in bed?"*

He turned back to Flieger with the question; *"They want to know; "Both of them at the same time or just one?"*

"I don't believe you! I'll ask myself!"

With a lot of hand signals and using his simplest English he got the message across to the young Senoritas and they pointed at one of them "tres", and both of them "cinco!"

"They're cheaper by the dozen!" Forrer laughed!

"I can't believe this. They've only been here for two weeks."

Forrer asked the little one with the long hair. She looked like she was 18 or 19, 20 maybe!

"QUANTOS ANOS *TIENES?"*

"QUINCE!"

He nearly choked on his beer. *"Frits,she's only fifteen. Let's wash up and split this place! If we get out of here now, we can make Tucson before dark. Tucson is supposed to be an exciting place!"*

"Why? I like it here!"

"I do too, but what are we gonna accomplish? Get shit-faced again? Fall tearfully in love again? Try once more to catch the clap? Come on!!! Let's get the hell away from this place! We wanted to see Mexico, right? Well, I saw it! Let's go!"

"But...."

"You got laid, so that should hold you over for a while!"

Wrong!!!!! That night in Tucson, a young pretty blond walked out on him and left him slobbering at the bar; *"I love her so much, boy oh boy, oh boy! And now she left....."*

It was the same old story....they had gotten along just great, danced intimately, laughed and drank and then he became too drunk and overly sentimental and told her that she was the greatest love of his life and she got turned off! It probably had happened to him a hundred times.

The next morning he couldn't even remember her name or her face. Actually there were two Fritses who should give up drinking so much!!!!

CHAPTER XIV
LUKE A.F.B

They arrived a day early at LUKE A.F.B. and that allowed them to get checked in and catch up on their sleep. They were looking forward to a quiet evening or two.

Then they'd be ready to dive headlong into their newest adventure.....flying the F-84 THUNDERJET!!!!

Things didn't quite work out that way......

An embarrassing situation existed on the base. All American Pilots were officers. Most all of the Foreign Pilots were Sergeants! They shouldn't be allowed in the Officers Club and the American Pilots weren't allowed in the N.C.O. Club! The brass really didn't know what to do with them....they're all pilots.....they fly on each others wings.... the Sergeant may be leading the 2nd Lieutenant......In the air, they're all the same, on the ground they're all the same too......except the rank the various countries assigned them.....what the hell do you do with all these foreign monkeys?????????

Some of the Non-Com pilots might be better pilots then the Officer-Pilots?.....Who knows????? They all qualified from the same training program, didn't they?????

A very, very awkward situation!!!!!

They are at least equals... ... NO??????

SOLUTION....AN F-84-CLUB!!!!!!

A separate club for the foreign students!!!!!!

Well... the F-84-Club was a blooming madhouse!!!!!! The Joint was a wall to wall madhouse of pilots on a drunk!!!!

The place was a madhouse like the world may never have known!!!

It was the grandest continuous party unknown to mankind!!!!!

Young, brand new pilots from all over the world congregated in the F-84 Club to brag about flying experiences, girls, play cards, sing songs, tell jokes while drinking beer incessantly!

It also provided a perfect setting for an old fashioned "HOMECOMING" party. Guys were reunited who hadn't seen one another for more than a year! Frits hadn't seen Jan de Ruiter since New York. Hadn't seen Henk Benninga since Bryan. New faces were introduced; it was a backslapping, handshaking congregation under the barracks roof that housed the "Club!"

Hours of catching up on each others "war stories" always included loads of beer and there seemed no end to the supply! It was a "BLAST!"

There were also Danish, Norwegian, Belgian and French faces from the past and it was good to see that so many had survived the "WASH-OUT" axe.

The "old-timers" that had already been at LUKE for 3-4 weeks bragged about that "fantastic" flying machine, the F-84. Flying it was a gas, but learning "gunnery" was the ultimate joy of life. The newcomers were in awe and couldn't wait to get their hands on that monster jet.

Their chance came soon enough. The next morning! Not that they flew it already the next morning, but they touched it, they sat in it, they got to study and to feel it. They could already develop the sense of being "ONE" with the plane.

Workdays started early at LUKE, before sun-up. They met their instructors the very first morning. Most of the American Pilots had gone on to NELLIS A.F.B. for F-86 "SABRE" training, so most instructors had only one or no American students, maybe 2 Frenchmen and another Foreigner.

Captain King, Frits' instructor, was older than most of the men, 35 maybe, and he ended up with 2 Americans, 2 Frenchmen and Frits. He had a lined, weather beaten face, like an outdoorsman, a farmer or fisherman and he was a very soft spoken, tough minded man. Frits figured he would not fool around with this man. No, he would do his darnedest to please this Captain. Remembering his T-33 solo and his botched-up Instrument exams, he decided right then and there he was going to excel in this plane, starting with all the theory that came with it.

When running through the "emergency procedures", they came out of his mouth without hesitation and when they had to sit in the sun drenched cockpits to learn the "blindfold" procedures, he sweated through practice after practice until it was second nature.

The "blindfold" check consisted of the pilot sitting in the cockpit blindfolded and being able to point to and touch each instrument, switch, lever and circuit breaker as the instructor called them out.

The major difference between the F-84 and the other planes that they had flown was: IT WAS A SINGLE SEATER! No room for an instructor! The first flight was immediately their first solo-flight.

So it was imperative that they knew every procedure, every knob, every switch like the back of their hand. They should be able to start the engine, switch on the radio and be ready to roll with their eyes closed.

There were some two hundred switches, gauges, instruments and circuit breakers in that cockpit and it all looked as dazzling as the inside of an airliner. They had to know the location of all of them "blindfolded" or they didn't fly the damned thing.

With the instructor on the wing, calling off the items, they pointed at every one of them.

"OXYGEN!, ARTIFICIAL HORIZON, BOMBSWITCH, RADIO SELECTOR, HYDRAULIC PRESSURE, ROCKET CIRCUIT BREAKER, LANDING GEAR, DIVE BRAKE.....ON AND ON AND ON!

There were no options. Every pilot had to know it!... . and know it fast!!! They didn't fool around at LUKE!

Then came the first flight. Student in the lead, instructor on his wing. "*Taxi out! Ask for line up*!" After "ground check", "*Ask for take-off*"...."*Give it the gun*!"....and the new pilot was flying the F-84! The two of them would climb to a designated area and altitude and the instructor would follow through a series of turns, rolls, loops, climbs and dives. Then from an altitude of 11,000 feet, step by step, they'd go through all the pre-landing checks and procedures and shoot a few landing approaches from 11,000′ down to 10,000′, simulating an actual ground approach. By then it was time for the "real McCoy!"

"Luke tower this is "Hob-Nob" flight on I.P. for landing, over!"

The most dangerous part of this plane was the final part of the pattern. If, in the final turn, the airspeed dropped too low, the plane simply dropped out of the air like a rock and exploded a few hundred yards short of the runway. LUKE lost a plane a week like that! By the same token, if proper speed was maintained, she flew like a bird and landed like a charm. Nothing unusual! As a matter of fact, the aileron booster, like power steering in a car, made it a real easy plane to handle.

The airport was the busiest Frits had ever flown out of and it was just astounding to see all those formations taking off and landing in close sequence. Somehow the tower fitted him and his instructor into the hectic pattern and with the instructor right on his wing, he went through the whole landing process until the main gear hit the runway and the instructor called out: *"Hob-Nob, going around!"* UP dive brakes, FULL POWER, take her off, UP gear, UP flaps.... 400 feet.... turn up to downwind... . and do it again.... .

Three times,,, No problem,,,

He was now an F-84 pilot!!!

As soon as all the students were checked out they went on their first formation flights, practicing high and low level battle formations.

Capt. King was a man after Frits' heart. When he said; "low" level, he meant <u>"LOW"</u>! The students would lose sight of him behind a cactus or a rock! If there had been streets and houses they would have been flying through the streets looking in the second story bedroom windows.

Cruising at 420 MPH on the deck gave them a fantastic sensation of speed. Frits loved it! He could do this forever and ever! What a way to make a living!

Capt. King seemed generally pleased with his flights performance although he could get very impatient with his French students, especially when they spoke French amongst themselves.

"It's no wonder you goddamn Frenchmen are so slow learning English! All you do is speak French amongst yourselves.

The whole nine yards

Around here all I wanna hear is English, you hear?" "Only English!"

A new aspect was taught at LUKE; "AIR-TO-AIR GUNNERY!"

After running through about a week or two of "strafing", " rocketing" and "bombing" as they had started at "WILLIAMS", they started Air to Air. First with cameras again.

A B-25 or an F-80 would tow a target up to 12,000 feet and would fly it in a straight line at 200 MPH. The target was made of a fine nylon mesh and measured 10′ by 30′ long. It was white with a red border and a red two-foot circle in the middle, the BULLSEYE! The target was suspended behind the tow plane with a long rope and it was held in a vertical position by a metal bar at the leading edge with rollers at each end. When the plane made it's take-off run, the bar would simply roll across the runway, dragging the target until it was lifted air-borne.

Up at the range, there was again a set pattern for approaching and shooting at the target. The main difference between Air to Air and Air to Ground was kinda obvious.

THIS TARGET MOVED!!!! AT 200 MPH!!!! Considering how much difficulty they had experienced learning to shoot at a target that stood still, they could well imagine what they were in for with a moving target. Consequently, the pattern also moved right along with the target, at 200 MPH!! In close formation the flight would come abreast with the target, doing exactly 200, but 4000′ higher and 4000′ out. That was called the "PERCH!" Like a parrot sitting on his "perch!" If the pilot would now dive straight down at his target, the target would flash by him like a racecar passing a bystander at 200 MPH! He couldn't possibly get a shot off! If the pilot went to fly behind it, he would have all the time in the world to fire, but he would only be able to see a vertical line in the air, the thickness of the nylon and he'd be shooting the tow-plane right up the butt!

There was a better way! From the perch, doing 200 MPH, the Flight leader would roll in steeply, putting the plane on it's wing while diving down at the same time. By making that maneuver he'd fall behind the target a little, but at the same time gain speed. Halfway to the target, he'd reverse direction and turn directly into the target, that was still slightly below him and slightly ahead. As he closed in on the target and put his "PIP" on the bullseye, his speed was approaching 400 MPH and his "TWO-MILL PIP" should become the same size as the bullseye at 1000 feet out. Then *"TATATATATATATA"*, for a few seconds he'd pull the trigger and immediately break off in order to avoid colliding with the target. He'd level his wings, slide over and behind the target, give it the gun and climb out ahead of the towplane.

He'd level off again on the perch, about even with the tow plane, 4000' above it and 4000' out, going at the same speed.

Theoretically, all his bullets hit the target! If he had been exactly 1000' out, doing exactly 400 mph, his "PIP" directly on the bullseye, not skidding or slipping through the air by using his rudders unnecessarily, yes then indeed his bullets should all be "HITS!" In reality, it all happened so fast that "*DAMN*", they were by the target and on the way back up before they even had a chance to line up with the target. Everything had to be so damn precise and correct, that it turned out to be one of the greatest challenges they had ever flown. It would have been difficult enough doing it with just ONE plane, but with the four of them in that continuous carousel it seemed impossible.

All sorte of things could go wrong. If a plane ended up on the perch too far ahead of the tow plane, his pass at the flag would be less then 100th of a second as the target flashed by him. If he ended up on the perch behind the target, he'd end up behind the tow plane, unable to get a shot off. Also, they could stay up on the perch too long trying to adjust their position because another plane would be calling...#...OFF, doing 500 MPH and an F-84 would be climbing right under their tail. The pattern had to keep moving. It was a continuous "rolling in", "reverse the turn", "shoot", "break-off", "power and pull up", "roll out", "listen for #....IN", roll in again and on and on and on. A rhythm was established and should be maintained like clockwork. "# 1, IN", "#1 OFF", immediately followed by "#2 IN", a few seconds and "#2 OFF", "#3 IN", etc.

Keeping track of everybody in the pattern and doing a good job of flying the pattern properly was a task all by it self. If any one of the 4 pilots was too slow, too fast, too far out or too close in, he'd screw up the pattern for the rest of them.

Many a time it all went so fast that the new pilot didn't even squeeze the trigger, not even for one second.

Practice man, practice!

That's why they first worked with cameras only. On the ground they could view the films and go over their mistakes and work on improving them in the safety of an armchair.

Then came the "LIVE" bullets!

The tips of the bullets were dipped in different colors and

they left a trace of the paint on the nylon fiber around the bullet hole. Let's say the Flightleader had "yellow", #2 had "purple", #3 green and #4 black.

Upon the return of the planes, the ground crew counted the bullets fired, let's say 100, then count the colored holes in the target, let's say 20 "yellow", then the pilot's score was 20%. In order to be considered "COMBAT-READY", a pilot had to have an average score of 6% or better. That low a requirement should have told them in advance, that it wasn't going to be easy and nobody expected any miracles. Most instructors fired around 30%, but they had become masters at it. Frits' scores stunk most of the time and he ended up with a mere 7% for Air to Air.

Thank God, his ground scores were much higher. As far as the Air to Air scores were concerned, his subconscious might have been at fault.

On the second Air to Air flight with "LIVE" ammo, he just called "#4 OFF" and stood nearly straight on his tail on his way up to the perch as a French voice screamed over the radio:

"OB NOB TWO, OB NOB FOUR, MEED-AIR COLISIOON!!!"
"OB-NOB TWO, OB-NOB FOUR, MEED-AIR-COLEESIOON!!!"

At the same time a flash reflected in Frits' rearview mirror and with a jerk he whipped the plane to his right and whipped it back around to his left, upside down, looking for the # 2 plane he was about to collide with!!!

"OB-NOB TWO, OB-NOB FOUR, MEED-AIR-COLLISION!!

"SHUT UP THREE!!!!! THIS IS FOUR, I'M NOT IN A COLLISION!!! "WHAT HAPPENED???

"OB-NOB TWO, OB-NOB FOUR, MEE-AI,,,"

"TA GEUILLE, TOI!!" "RETOURNEZ A BASE!!!!

He had to shut the stupid bastard up, he was hogging the airways. At the same time he saw a flash in the desert, cut power, hit the dive brake and dove toward the flash, standing straight on his nose! Another flash!!!, not far from the first...*"GODDAMN!!"*

He said to himself:***"Those have to be #1 and #2!!!"*** He hit the radio button, hollering loudly:

"Mayday, mayday!!!!" "This is Hob-Nob 4, we've had a mid-air collision! How do you read? Over"

"Hob-Nob 4, this is Angel leader, what is your position ? Over"

"This is Hob-Nob 4 going down to the deck, about 50 miles

west-southwest of Gila-Bend, there were two crashes in the desert, over!"

"Roger, Hob-Nob 4, I've called the base. Chopper on the way! Do you see chutes and try to give your exact position, over!"

"Roger Angel lead, I'm passing through 5000 feet and no parachu....wait,,,,, yes one,,,,just one,,,, just touched down,,,, he's o,,,, another one,,,, we've got another one,,, I'm following it down, over!"

"Roger Hob-Nob four, keep 'em in sight, circle at 2000 feet and try to pinpoint your position, over!"

"Roger, Angel lead, #2 chute landed hard! Oscilated, pilot not moving! #1 chute o.k., pilot covered up! I'm due north of the highest peak of the Mohawk Mountains, about 15 miles in a standard left turn at 2000, over!"

"Good work, number 4, where's #3? Over"

"Sent him home,,,, got you in sight!! I'm at 10 o'clock low from you, 5 miles out, over!"

"Roger, Hob-Nob 4, this is Rosebud leader, got you in sight! Angel lead, I'm at 6000' near the crash site, going in! Hob-Nob 4, keep on looking out for Angel lead, over!"

"Roger, this is Hob-Nob 4!"

"Rosebud leader, this is Angel lead. Got both of you in sight. Will direct the chopper. Hob-Nob 4, return to base! Good work, out!"

"Roger, Hob-Nob 4 returning to base!"

Close to the base, approaching I.P., he spotted Hob-Nob 3, wandering over the desert! After trying to reach him on different radio channels, he joined up with him, got him to fall back on his left wing and "easy-go" brought them home. He wondered what the clown had been doing all this time?

On the ground the whole story unfolded. The reason why the Frenchman called out; "HOB-NOB TWO, HOB-NOB FOUR!" was logical in his mind. He was #3 and he didn't collide with anybody and it couldn't be #1, because instructors were infallible. So much for that!

The other instructors had immediately sent their flights home by themselves and set out looking for survivors, not really knowing where to look. Frits, being the number 4 man, had been so busy flying formation to the target and flying the pattern to the target, that he knew only vaguely what his position was.

The mountain range to the south was the only thing that stood out in that area. It was the only reference he had, but it worked!

The actual collision was an intriguing one! #1 was coming in on a pretty normal approach, about a 25% angle to the target with his eyes locked on the bulls eye.

#2 had pulled too far forward on the perch by pulling up too steeply or giving too much throttle and rolled in too fast behind #1, coming in at a 50-60 degree angle to the target. He couldn't see #1 under his belly and #1 could possibly be expected to look up and behind him while barreling in on the target at 400 MPH. As they approached the 1000 foot mark at the same time, #1's right wing crashed into #2's belly, causing the explosion that Frits saw reflected in his mirror.

#2 pulled up the ejection handle, blew off the canopy, pulled the trigger and shot himself free of the plane. After undoing his ejection seat, he pulled the handle on his parachute and drifted down to earth, watching his plane explode in the desert below him. At that point, he was still not aware that he'd been involved in a collision, he still thought that his engine had exploded! He landed nicely and, according to the book, he covered himself with his parachute to shield him from the sun until rescued.

The instructor didn't fare so well! All of a sudden, just as he pulled the trigger, there was a terrible crash! He didn't know what hit him! His right wing snapped off like a matchstick! He snapped upside down in a violent spin with so many negative "G's" that he couldn't get his hand to the "trigger" of the ejection seat after he blew off his canopy. Besides that, because of his violent spinning motion with just one wing, his canopy didn't blow straight "UP", (or "DOWN", actually), it kinda blew sideways, hitting him in the helmet so hard that it cracked his jaw and nearly knocked him out!

Being the veteran pilot that he was, he kept his wits about him and managed to undo his safety belt, then with both hands he shoved the stick forward as hard as he could. Since that he was in an upside down position, he pitched the nose upward and the plane threw him down and clear of the tail! He pulled his parachute cord and floated to earth.

As fate would have it, the hot air rising off the desert floor made his parachute oscillate, lose lift and dropped him down to the ground so hard, it knocked him out!

Two days later, he was back!

#3 was somehow exonerated, but assigned to another instructor.

#2 resigned from flying and went home.

Capt. King stated; " *I'm gonna need another student. I think I'll ask for a Dutchman,O.K.?"*

His jaw was cracked, not broken and it didn't need sewing or stitching, but it would be painful for a while.

With a crooked grin, he joked; *"I'm now on an all beer and coffee diet!"*

In spite of the fact that he received a lot of compliments about his handling of the emergency, Frits was not at ease in the pattern thereafter. He found himself wanting to keep a visual check on everybody else and as a result he hardly ever made a perfect pass, in which he could pull that trigger at the target for at least two seconds. Most of the time, he knew he would miss as he pulled the trigger anyway!

Nobody, no instructor, no wingmen, no company commander bugged him about it, though! The captain said nothing and neither did anybody else. They must have known that there was something he had to outgrow!

It didn't influence the rest of flying any. When they went on "SEARCH AND DESTROY" missions, Korean style, attacking a convoy in the mountains, he was in his glory!

Convoys of old Army trucks and tanks were driven into the mountains, disabled, (so the Indians wouldn't remove them) and then dispersed and camouflaged the way the North Koreans went about it. It would be the flight's mission to find the convoy and then go about the business of putting rockets in the back of trucks and smash bombs into old beat-up tanks. It was a glorious time for young fly boys, especially since nobody shot back at them!

It was all very suddenly interrupted by 18 days of Christmas leave!

CHAPTER XV
CHRISTMAS 1953

"THE DUTCH CONSUL IN LOS ANGELES HAS INVITED ALL DUTCH PILOTS TO A NEW YEARS EVE PARTY IN LOS ANGELES WITH ACCOMMODATIONS AVAILABLE AT THE AIR FORCE RESERVE HEADQUARTERS AT THE LONG-BEACH MUNICIPAL AIRFORCE BASE. VISITS TO T.V. AND MOVIE STUDIOS WILL BE ARRANGED AND CHRISTMAS DINNER WITH PROMINENT DUTCH RESIDENTS IN THE L.A. AREA WILL BE ATTEMPTED."

"What do you say?" "Are we going?" "Yeah, let's go!" "Are you driving?" "Bus?" "Hitch-hiking?" "How much are bus tickets?"

"Can we get the Air Force to hop us over?" "Wanna take some F-84's ?" "I've never been to the west coast!" "Hollywood, here I come! Step aside Clark Gable! Get ready, Doris Day!"

Well,,,,they went!

"We'll get quarters at Long Beach first and we'll take it from there. Tomorrow night we'll meet in the N.C.O. club for cocktails!" "Alright?" "Alright!"

Frits went hitchhiking by himself. He was optimistic about the prospects and he told the others; *"I might beat you to it and you have no idea what you can run into! Besides, I'll save $14.50!"*

It took longer than he had planned, but that didn't bother him that much. He was in time for cocktails at 5! He had slept in a truck cab the night before, because the trucker went to sleep on his bunk in the rig.

The WAC Corporal, "Cpl. ANDIAMO" according to her nametag told him as he checked in; *"You're late, sir! The other Frits and the rest are already here!"*

"I'm Frits, not "sir" and what's your first name?"

"Mary-Ann,Mary hyphen Ann!"

"How do you spell hyphen?"

For a moment she looked dumbfounded, then she burst out laughing showing all of her pretty white teeth. She looked like Elizabeth Taylor with blue eyes!

"If I weren't so damn old, I'd invite you for cocktails too! But I'm old enough to be your father!"

"How old do you think I am?"

"Oh, I'd say,,,about 19, but you shouldn't hang out with over-the-hill guys like me!"

"In that case you have bunk #2 in room 12, buiding "D".

At least she smiled as she said it!

"Maybe I'm not too old after all!" he whispered as he walked out of the office. *"What a charmer*!", he said to himself.

At the bar, the gang met, partied and planned. Someone had the schedule. Two days from today: Red Skelton show rehearsal!

Lunch at WARNER BROS studios. Visit to REPUBLIC studios and bus tour of Hollywood. Cocktail party at REPUBLIC Director Weintraub's house. Christmas day, we'll be picked up by Dutch families over here. We'll need a list of interested guys.

New Years Eve dance in GALA dress uniform, guests of the Consul General! Again, need list of interested people!

"Gala dress uniform? The blues I own are 2 years old and certainly not dress blues!"

"Hell, we'll find something else to do! Put me down for Hollywood and Christmas. Who needs another beer?"

Frits walked over to the bar and nearly bumped into somebody! *"Oh, my! Mary-hyphen-Ann!!!!! My, do you look different!!!"*

"Better or worse?"

"You couldn't look better, but civvies become you and now you look even more like Elizabeth Taylor with your hair down!"

"What do you mean; I couldn't look any better?"

"You looked so good this afternoon, you could hardly improve upon it! What do you drink?"

"You guys are all a bunch of flirts, do you know that?"

"Do you like that? And what do you drink?"

"A beer please!"

"Millers o.k.?"

"Fine!"

"Here! Stand here a minute. I've got to deliver these! I'll be right back. Don't move!"

When he made it back, she had found a stool so he stood next to her. "*Wow, does she look fine*!" He was thinking, "*What a doll!*"

"Why are you guys pilots and not officers? Why couldn't we put you in the B.O.Q.'s?"

"*Well, hell, here I go again!*" He told her the same story he had told so often and she reacted; "*I don't belong here either!*"

"What do you mean?"

"As a corporal, l don't belong in an N.C.O.. club, but being that most everyone is gone on holiday leave, the other facilities are closed and we're heaped together in here! Isn't that cosy?"

"I'll drink to that and to Mary-hyphen!"

"*Cut that out*!" An elbow poked into his ribcage for emphasis and he knew; "*I'm in like Flynn*!"

The Air Force Reserve motor pool didn't have much available for transportation, nor much personnel to run it, but they did provide a bus to the HUNTINGTON RAILWAY MUSEUM in Huntington Beach and back in time for lunch. Later the bus took them to the bus station for those who wanted to go and see L.A..

The boys took them up on both offers, also checking out the Long Beach pier and the cold Pacific Ocean. That was a FIRST for most of them.

At night back to the N.C.O. Club for free food and Mary-Ann.

She was just 19 and built the way GOD originally must have designed a woman. Fine, fine shape! Clear skin and smiling eyes framed by all that pitch black hair. Beautiful! She was also a sport, although she drew a line. He had known her for just two nights and he knew that she would smooch, but not hit the sack with him!!! "*IT'LL COME!*", he told himself confidently.

The Hollywood trip was hysterical! A bus was provided and the first stop was "RED SKELTON".

They were required to dress in uniform and the 17 of them, NETHERLANDS on their sleeves, two wings on their chest, received as much attention as the T.V. Stars. The rehearsal before a live audience was the funniest thing they had ever seen. Because the actors would screw up regularly and burst out laughing themselves, they kept the audience in stitches. It was a jungle scene with SKELTON as an overweight, undersized TARZAN in a leopard loincloth, EVA GABOR as JANE in a very skimpy outfit and JOHN WAYNE dropping in on them in a western outfit, asking; *"WHICH WAY DID THEY GO?"*

At lunch they freely mingled with the stars at the cafeteria and Frits felt very smitten with Eva Gabor.

They had to move on though. Bus tour of Hollywood, celebrity homes and oil wells on a school ground! Then off to Republic Studios! What connection, if any, there might have been between Republic Aviation, the builders of the F-84 and Republic Studios, they never discovered. A WESTERN was being filmed, starring LIZ SCOTT (husky voice and all) and JOHN PAYNE.

Watching a movie being shot was an enormous thrill for them and getting to meet the stars in person was a experience of a lifetime. Getting to shake hands with Liz Scott before she climbed into her red Olds convertible made him forget all about Eva Gabor in a heartbeat. There was something familiar about her red car, as if he'd seen it before, but he couldn't place it.

The cocktail party at the Director's house was elaborate. The setting at poolside, overlooking L.A. was magnificent!

Every type of booze was available and every kind of food imaginable! The boys managed to make complete asses of themselves, Frits included, by trying to eat everything in sight, drinking too much of too many different kinds of booze and most of them were pretty well loaded within the hour. Within two hours, the bus was on the way back to Long Beach. They had a lot of laughs and the whole day was a great experience, but with a little more moderation the evening could have lasted a lot longer and been a lot more memorable!

When they got back to the base, the club was deserted and Mary-A was nowhere in sight. Actually, nobody needed any more to drink, so they hit the sack! So much for their one chance in a lifetime to make out with a movie star!

Two days later was Christmas. A Christmas dinner was being organized in the N.C.O. Club for all military personnel that stayed on base and Frits was tempted to stay, but he was on the list of people to be picked up and a DR. BEENMAN arrived at 10 AM to collect three of them. *"BRING BATHING SUITS!"*, the Doctor suggested and they were on the way to his home. It was a nice, cool but sunny day up in Beverly Hills where the Doctor and his American born wife made their home. From their pool patio, they could overlook all of Los Angeles and way out into the Pacific.

The temperature lingered in the seventies for a few hours, allowing them to lounge about and swim for a while.

According to their host, the smog was unusually light that day and they could see the Wilson Observatory toward the east.

The BEENMANS had invited a few young ladies to dinner as well and it made for a very nice, polite, but subdued gathering.

Everything was so dainty, so proper, so perfect, that Frits didn't even feel like bursting out in song or telling a few rowdy jokes.

It was nice though. Tea was served by the pool, sherry before dinner and wine with dinner. Nice little gifts were opened afterwards. Nice touch! Very gracious people. An experience of a lifetime, Christmas in a bathing suit up in snazzy Beverly Hills. He wouldn't have missed it for the world.

When at 8:30 it was suggested that they might be getting tired and might like to get back, they politely accepted "so the doctor could be back home at a reasonable hour!" It had been fun!

"I'd Know, I'd Go From Rags To Riches!" blared in their ears as they walked into the Club, *"If You Would Only Say You Care!"* he joined Tony Bennett full force.

"My Clothes May All Be Torn And Tattered, But In My Heart I'd Be A King!,,, "And Though My Pockets May Be Empty, I'd Be A Millionaire". He cruised all around the bar looking for Mary-A... . Not in sight! The accordion door to the banquet room was still open, showing the empty Christmas-dinner-tables. He peeked inside and there she was!!!! Sitting with her back to him, she was talking to another gal in an evening dress. He snuck behind her, covered her eyes and whispered: *"You'd Be My Everything!"*

She swung around and he stopped, baffled! She was crying, mascara streaked down her face. Somehow, it made her look girlishly beautiful.

"OH, FRITS!" she cried. She grabbed him around the neck

and sobbed, her whole body shaking and tears running down his suit jacket.

"*What's with her?*" he mouthed silently at the girl across the table. She made a drinking motion with her hand and shrugged her shoulders. Frits grinned. He took Mary-Ann's face between his hands and gingerly kissed the tears off her cheeks. He put her head back to his chest, produced some greens with his left hand and holding them toward the other girl, he whispered; *"A beer, a soda and whatever you want!"*

"*That's alright. I got it!*" She dashed off.

Frits went back to tears and cheeks and although the taste of mascara mixed with powder and tears was not the greatest tasting combination in the world, he thoroughly enjoyed it.

Mary-Ann calmed down some and after "JACKIE", the other WAC, had deposited the drinks and left, Mary-Ann started to explain tearfully; *"This is the first time in my life I haven't spent Christmas with my family and I started to feel so lonesome!"* More tears, more kisses and a ruined face that held a natural glow without makeup.

After some soda and some dancing, she changed from a sentimental drunk to "just happily sloshed!" They whirled to the jukebox till closing time. That night they slept in one another's arms with most of their clothes on, celibate but content!

On Saturday night the boys explored L.A. and wound up in the "OLD HEIDELBERG" with a lot of "HOOMPAHPAH", "STEINS" of beer and "BRATWURST". Plenty of "WALTZES" and "POLKAS" filled the air.

When they played some old standards like: "*Es Geht Alles Vorüber, Es Geht Alles Vorbei!*" and "*Wer Soll Das Bezahlen?*" Frits helped the band by singing at the top of his lungs and he certainly helped the group of young guys to catch a lot of attention. It didn't take long before the customary "*WHERE YOU GUYS FROM?*" started to be asked, usually followed by "*OH, I'VE BEEN THERE!*" or "*I ALWAYS WANTED TO GO THERE!*"

It was a surefire way of getting involved with the rest of the crowd and it lead to more "*TRINKINK AND SINGINK*" and rounds of drinks all around. Their intent had only been to "check out the joint", but they hung around till closing time with an invitation to lunch and a soccer match the next day. They would play on the SWISS team against ITALY. Not nationally of course, but with a local soccer league.

That sounded like fun. Nobody in the group had offered them a bed to sleep in, so after one more night club stop, they had to start looking for lodging. A San Diego Marine at the bar had suggested the Salvation Army; *"THEY'RE CLEAN AND CHARGE NOTHING, BUT IF YOU WANT BREAKFAST YOU'VE GOT TO ATTEND THEIR GODDAMN CHURCH SERVICE!"*

Somehow that sounded sacrilegious, but they went anyway and six of them slept on cots in the Salvation Army. Most gracious people! Even if the singing at the morning service only mildly offended their headaches.

The lunch was stupendous! Swiss-German food heaped in stacks so high, that it resembled the Alps. They gulped it down with so much beer that they needed a siesta more than a soccer match.

The match was fun though! Most guys were sadly out of condition, but the Swiss team lead by the score of 7-0, so in the second half Frits became Italian and tried to save the team, but the Swiss ended up winning by the score of 10-4. The important thing was, the match had been a real crowd pleaser and it resulted again in a true "FEST" with lots of "*WEIN, WEIB UND GESANG*!"

The Dutch guys asked in amazement; '*Where did you where learn all those German songs*?"

Facetious as he could be, Frits answered; *"I grew up under German occupation! Where were you?"*

Of course, with the exception of Frits Flieger, so had the others, but not with the benefit of having a bar at home, crowded with singing KRAUTS.

He did have an edge and enjoyed showing it off.

Some friendly folks drove them to the bus stop for Long Beach and they happily sang their way back to base.

As the door to the N.C.O. Club opened, the "FOUR ACES" greeted them with "*Heart Of My Heart*!" but his heartthrob Mary-A was nowhere in sight. *"I Love That Melody....It's My Favorite*!"

His "favorite" was not at the bar, not crying in the backroom but "Jackie" noticed him, jumped up from the bar and ran towards him. "*She just left, I'll go get her*!" Gone!

Within two minutes she was back. "*She'll be right here*!"

Just as the second beer arrived and JO STAFFORD began singing; "*Take Me In Your Arms And Never Let Me Go*!", she walked in. Frits met her halfway to the bar and swung her around; "*Whisper To Me Softly While The Moon Is Low*!" She was a great jitterbugger, "Amsterdam" style with a bit more flair!

"*Make Love To Me!*" Jo Stafford was crooning.

"*I'm ready, I'm ready*!" Frits was hollering as they swayed across the floor.

"*I'm not, I'm not*!" she laughed. My, was she pretty when she laughed! "*When You're Near, So Help Me Dear, Chills Run Down My Spine!*"

"*True! True!*" it sounded like he was coo-ing like a dove!

"*Don't You Know I Love You So?*" Stafford and Frits!

"*No, I don't, I don't*!" Mary-A came back!

"I Wont Be Happy Until You're Mine!"

"I can't believe that, you flirty Dutchman!"

Much later, danced out, she got serious, too serious! "*I like you a lot Frits, maybe too much! But in a few days you'll be gone, I'll still be here and before long you'll be back in Holland and I'll never see you again!*"

"Who knows,,,,,?"

"*Be quiet! I was talking.*" She tapped him on the nose with her fore finger, like giving him a minor spanking, " *So I hope to enjoy your company as much as possible while you're here, but no sex! If that makes you mad and you want to chase some of the other girl, I'll understand!*"

"*You're not gonna seduce me?'*

"Ha ha ha, I'm sure you'll hold me to my word and fight me off, right?"

"Right, that's exactly what I was gonna say! What would you expect me to do? Give in? Like the guy who said; "SO HELP ME, I'LL RAPE YOU!" and she answered: "SO RAPE ME, I'LL HELP YOU!"

She couldn't stop laughing,,,,; "*You're a nut Frits, but at least you're a lovable nut*!" She kissed him tenderly and he slipped his hand up to her boob, whispering between kisses "*You're gonna seduce me yet!*"

When New Year's Eve came around, Frits had decided to skip the "BALL" and stay at the party on the base. Most of the guys did go to the "GALA" and wished him a "BALL" anyway.

According to the posted agenda, the party on the base was going to be a good one with just one big question; "how good is the band?"

They turned out to be just fine and Mary-A had surprised

him by paying for him as HER guest! "*This is the last time of the year that I can spoil you and besides, you treat me all the time*!"

The food, the music, the noisemakers, the California Champagne, all was fine if not great. The champagne was part of the deal, but Frits bought his beer on the side anyway!

"*Champagne leaves my throat so dry!*" was his excuse.

When the band struck up "*La Vie En Rose*", he got up behind the mike and sang it in French, English and French again and it was well received.

"*A little touch of culture couldn't hurt*!" he told the Mess President who complimented him.

Mary-Ann went soggy on him; "*That was so beautiful! I felt like you were singing only for me*!" For a minute he thought she was going to cry again!

"You couldn't understand a word I was singing!"

"Yes, but I could just feel the meaning! You've got to sing that to me again,,,,in private!"

At midnight she pledged; "*I'm gonna ask for a European tour of duty! I'm gonna catch up to you yet*."

"I'll drink to that!"

They slept together again. No clothes,,,,but no sex either!

In a way he didn't mind. It wouldn't have been right, she had too much champagne. She might not even have remembered it.

On New Years Day, he was nearly subjected to that ridiculous American ritual "THE ROSE BOWL!"

Mary-A's team "INDIANA" was in it and she watched the entire game in the T.V. room with all the other "football nuts".

It was an exciting and tense game! As a matter of fact, it was so exciting and so tense that Mary-A had chewed up all her beautiful fingernails, right to the bone! She was nearly hysterical! "*Look what I did to my nails!*", She screamed, "*Look, Look! Oh,,, this is awful, just awful!*"

"Oh well, it proves you had good time, right?"

She could have hit him!

The next day, a working day, she was back in uniform, her hair the same way as when he met her. He fought off a serious pang in his heart! He really liked that girl!

It was hard saying "farewell". For both of them.

Back at LUKE, the military training machine was cranked up in such a hurry that it might as well not have stopped. The planes roared up again from before sun-up and Frits was scheduled for one of the early flights, which nearly became his last!

It was going to be a "SEARCH AND DESTROY" mission with everything loaded and attached; 2 bombs, 16 rockets and all 6 guns loaded.

He was flying # 4 again, damnit, so he would have to work hard again. He ran his exterior check, crawled in the cockpit and with the help of a crew chief, got himself buckled in. Being a bit of a wise guy, he didn't think he needed to go by the checklist and did it all by heart.

When #1 gave the signal, making circles in the sky with his hand, *"START ENGINES"*, he started up, checked the gauges and switched on the radio. He expected to hear #1 on "E" channel, the ground-control frequency, *"LUKE GROUND, THIS IS HOB NOB FLIGHT, RADIO CHECK, OVER!"*

"YOU'RE FIVE BY FIVE!" was then the answer from the tower, but today there was nothing.. *"GENERATOR AGAIN?"* was his first though. "NO, NO,!", everything worked like a charm. He turned up the volume, nothing! He switched to "F" anticipating the call *"HOB NOB FLIGHT, CHECK IN!"* and then *"HOB NOB TWO"* and *HOB NOB THREE!"*, where after he would call in. "NOTHING"! Circuit breakers.....all in.... everything's O.K.

He saw number 1 taxiing out... *HOTDAMN*... he would have to abort the mission...."*LIKE HELL!"*...he fell in right behind #3, while still checking switches and button and circuit breakers.

"DAMN, DAMN, DAMN!" He knew he would have to turn back, but yet he hung right in there in perfect taxi formation. They were closing in on the runway when he noticed that his oxygen-blinker didn't blink. "WHY NOT?" He checked his connection....*"HOTDAMN"* ..it wasn't plugged in...and neither was his radio cord. He plugged them in and nearly busted his eardrums. He had turned the volume to the max and it just about blew his brains out. *"OH WHEW"*, he had lucked out!

Just then..."*HOB NOB FOUR, DO YOU READ?"*

"FIVE BY FIVE! HOB NOB 4." He acted as if nothing had happened! It was just in time, though!

They were behind a flight that was just lining up for take-off, well... . they weren't too organized either. Their #4 man was trailing behind and made such a screeching, sharp turn onto the runway, that two of his rockets fell off and rolled along the taxiway into the dirt.

Frits laughed out loud; *"What a dumb fuckup! At least I'm not the only one that screws up today!"*

Numbers 1 and 2 pulled onto the runway followed by #3 with Frits on his right wing.

When 3# drew circles in the air with his finger, Frits shoved open the throttle while #'s 1 and 2 were leaving in a plume of smoke. #3 put his hand next to his helmet and suddenly threw his hand forward, releasing his brakes at the same time. Frits was right with him! The two planes raced together toward the end of the runway....60 MPH,....70,....80,....100,...120...

140...150... # 3 lifts his nose....Frits pulled back on his stick as well.....*IT'S STUCK...HOTDAMN*.....HE PULLED.....SHOOK.... IT STAYED LOCKED SOLID....The end of the runway is fast racing toward him....at 160 MPH.....it's too late to stop......and he can't get it to fly....he's gonna crash on top of his bombs and rockets and blow himself to smittereens.....No way to stop...cut throttle...eject armament.....hit brakes and he'll hit the desert at 100 MPH.....within seconds... ..just a few have gone by....and he has just a few more to live......

He looked down briefly....."*HIS LOCK!"....HIS CONTROL LOCK!*

He slapped his left hand down and snapped up the lock handle while his right hand pulled him off the ground as the end of the runway disappeared underneath him.

He popped up like a helicopter, passing his element leader, #3, who was already at 200'. With his heart pounding in his chest he settled down on the right wing of #3. "*Good God, that was close! That was too damn close*!" He had seen the pearly gates there for a second. "*What a stupid thing to do*!" For a year and a half now, before every flight, he had checked the controls for "freedom of movement"! On every flight, the same routine..."pull back that stick, look over your shoulders at the elevator.

Kick those rudders,,,far left,,,far right,,,,lay that stick all the way to one side,,,now the other,,,,,look at the ailerons, make sure one is up while the other is down,,,!" Every day of his flying life!!! Not today!!!! He hadn't set the trim either.

He hadn't had his head down in that cockpit at all! "*DAMN,*

DAMN, DAMN!" That had nearly been *"GOODBYE FRITSIE!!!"*

If he had lowered his head into the cockpit just once he would have noticed his oxygen hose unplugged, ditto for his radio line, he would have noticed that the oxygen blinker on his right wasn't blinking, he would have noticed that his trim wasn't set for "heavy load" take- off! *"DAMN, DAMN, DAMN!"* He was talking to himself as he adjusted to "low-level-battle formation".

"I've got to get my head out of my ass fast or I'll be a statistic, dammit!"

His heartbeat settled down to normal, his breathing leveled off and he concentrated on the flight at hand. There was plenty to concentrate on. They were flying so low, that from time to time he lost sight of his wingmen by flying between 30' rocks.

That was fun the old fashioned way! When at a 1000' at 420 MPH, the F-84 moved along nicely, at ground level at 420 MPH, the ground would scoot by them so rapidly, they could barely see it!

Obviously, they didn't concentrate on objects close by. Objects a quarter of a mile in front of them determined corrections in altitude or direction. It was exhilarating as all hell!

It all happened so fast, they couldn't afford to blink an eye or they might hit something. Supposedly, all this was to avoid "radar detection!" Their mission was: "SEARCH AND DESTROY!". Somewhere in the mountains was a tunnel in which a "KOREAN AMMO CACHE" was hidden and their job was: "PUT A BOMB OR TWO IN THE TUNNEL!". When the tunnel was spotted, they joined up in "tight" formation and rolled in, one after the other. *"BOMBSWITCH ON!, SIGHT ON!"*, everything working, "#4 IN"!. A hole in the bottom of a good sized hill (or small mountain) was the tunnel.

Diving down on a bombing run was nothing! Releasing the bomb just in time wasn't hard either. Putting the bomb right in the hole might also not have been a real problem, but doing all of those things and pulling out in time to keep from clobbering into the mountain,,,,,, THAT WAS TRICKY!!! To prove the point, airplane parts were scattered all over the mountain, showing that it had been done before! "Clobbered" that is!

"JEEZ, WHAT A THOUGHT!"

As he rolled in, he saw #3's bomb burst about 300 feet above the opening. *"PULLED UP TOO SOON, YOU DUMB BASTARD*!" he was thinking as he lined up and kept his PIP

500 pounders

on the hole.

"STEAD. . . . STEADY. . . PUSH!!". . . a little thumb pressure. . . . "bomb" gone!!! *"PULL UP HARD!!!"* His "G"-suit blew up hard against his belly and legs, his mask was being dragged down his face and it felt like his eyeballs were laying on his cheeks.

He must have been pulling 6 "G"s!!!

"#4, DIRECT HIT, PULL-OUT TOO DAMN CLOSE! #1 IN!"

Well, he got his bomb in, but according to Hob Nob Lead, his pull-out was too close.

Up on the perch, he relaxed and watched #1 pulling off.

His bomb was "right on!" and #2 was barrelling down toward the hole at the base of the mountain.

"This time I'm gonna do it differently!" he said to himself as he heard "2 OFF" and "3 IN". This time, *"4 IN, #3 200 FEET HIGH!"*, he dove steeper to the ground, aiming in front of the tunnel ! When he was way down, he brought the nose up, put the PIP right on the tunnel and gave full power. He hit the bomb release button and pulled up hard!

"#4 DIRECT HIT! GOOD SHOW!" That was #1 up on the perch, starting a slow circle so #2, #3 and #4 could join up with him.

Frits grinned to himself, he had seen P-51's, P-47's , Spitfires and Mosquitos do this type of thing day after day during the World War II back in Holland! He figured that coming in low might get the bomb to hit the ground in front of the hole, but by having more speed and a flatter trajectory, the bomb could skip across the ground in only one direction; INTO THE TUNNEL! It had worked! Also, by flying more level than at the 45 degree angle, he could pull up from the target with a lot more clearance. The fact that the instructor had not commented meant that it was O.K.!

Meanwhile the flight was off to find a convoy. Indians might have stripped tires and engines from some of the disabled vehicles, they might also be "slightly" shot up already, but they should still resemble a North Korean convoy on the warpath.

"HOB NOB, ARM ROCKETS, TWO AT THE TIME!" "LEAD" must have spotted the convoy. The rocket control knob was like a timer; set it on 1, only one rocket fired. Set on 2, it would fire one rocket from under each wing. Set it on 32 and *PFFFFFFFFFFFFFT* all rockets would fire in rapid sequence, alternating from wing to wing. All of them would be gone in about 5 seconds , rushing to the target like one big smoking fireball! They didn't make much noise, just *"PFFFFFFFFFT"* and they were gone.

Having 5 inch rockets that day, they only carried 16, instead of 32 "2.75" inch rockets and with their timer set on "2" they would make 8 rocket passes. That would be a lot of "G" pulling and all the "push-ups" from the past would be put to the test. With gunnery runs to come yet, their arms would feel like they'd come out of their sockets. Lifting beer cans and dancing over the last 20 days probably hadn't significantly improved his arm strength any… . what the heck!

"CONVOY AT 3 O'CLOCK LOW, RIGHT HAND PATTERN. TRUCKS FIRST, I TAKE LAST, #2 TAKE SECOND FROM THE END, ETC.! #1 IN!"

Since the planes were stacked on his right wing, #1 pulled up and kinda rolled over the top of his three wingmen and while upside down, pulled his nose down toward the target, rolled out and set his sights on the last truck. "#2 IN!" He pulled up and over,,,,,"#3 IN!",,,,,he pulled up and rolled over and when it was Frits' turn all he had to do was roll over, but he was so caught up in all the excitement he pulled up too and then over!

It felt like a million dollars! He had forgotten about his "near crash" just 30 minutes ago, he was too busy shooting the shit out of the "Koreans".

He was already "on target" when #3 fired.

"I'M TOO CLOSE!", he thought. He figured that in his enthusiasm he might have pulled too tight a turn.

"JUST HOLD OFF!" He waited, the "pip" lined up with the fourth truck and the moment he heard: "#3 OFF!" he let go; *PFFT, PFFFT*!" In a split second they were on their way and exploded underneath the canvas in the back of the truck. "#4 OFF!"

Reset timer to "4", so next trip he would fire "3 and 4" rockets.

"SAME TARGETS IN!. #1"

Same truck! This time he gave #3 more room. #3 wasn't pulling a very tight pattern. Nothing he could do about it!

On target,,,*PFFFT, PFFFT*! It was fun to watch those rockets race ahead of him toward the the truck and explode!

Nobody had to call out the previous man's hits because they could see the impact for themselves. Too bad the gun camera couldn't pick that up. He'd have something to brag about.

TANKS, UP AHEAD OF TRUCKS. #1 AND 2, LAST TANK. #3 AND 4, "TANKS, UP THE NEXT ONE!"

Where the hell did he see tanks? Where is he anyway?

A puff of smoke!!!! *"OH YEA!" "OH GOD, WE'RE WAY THE HELL OUT! HOPE HE SEES US!"*

"3 IN!" "DAMN,,,3'S MAKING THE LONGEST DAMN PASS INTO THE TARGET!!!! I WONDER IF HE EVEN SEES THE TANKS? I'D BETTER CALL!!" "HOB NOB 3 AND HOB NOB 4 ON LONG FINAL!"

"GOT YOU IN SIGHT. TIGHTEN IT UP 3!"

"THREE OFF!" He mustn'd have fired... . no puffs!

Frits had maintained some altitude and now dove down to the tank. "#4 IN!". It was a bit redundant, but he'd rather let everyone know his position. *"DAMN*!", those tanks were well camouflaged... "Fire!"... he squeezed off two rockets... . *"DAMN!", "BAD SHOT*!" They were off to the right. He shouldn't have made that last second correction! "4 OFF!"

Next trip, he would know where to look for that tank.

"ONE IN! 4 ROCKETS, SAME TARGET!"

Timer to "8". *"WE'RE RUNNING SHORT OF FUEL."* Frits

figured. Capt. King was speeding things up.

Three more passes with 4 rockets each and the tanks were pretty well demolished. Frits figured he had gotten 5 or 6 solid hits. Not great, 6 out of 16, but not bad either.

"HOB NOB, CHARGE GUNS!" Hob Nob Lead was wiggling his wings calling for "join-up!"

"GUN SWITCHES ON, LEFT PATTERN. JEEPS, 1,2,3 AND 4 FROM THE END OF THE COLUMN. JUST TWO PASSES, FIRE LIKE HELL! ONE IN!"

This was exciting, adrenaline pumping,, *"FOUR IN!"*

This time there was no mountain to worry about,,,, he could fire till he damn near hit the jeep with his plane.

"Keep that "PIP" on target, keep pushing forward on that stick,,,keep squeezing that trigger! With the increasing speed the nose wanted to come up and it required more and more forward pressure to keep the sight down on the target. *TATATATATATATATATATATATATATATATATA,* He blew the jeep to smittereens, parts flying all over and the plane felt as if he'd put the brakes on with the recoil of the 6 "point 50's".

"YOOHOO, FOUR OFF!" He yelled like an excited cowboy!

That was "live" action! One last pass! He'd imitate the wartime "JABO'S" again. The instructor might not be looking back at him after his own last pass!

"SAME TARGET, ONE IN!"

Some of the jeeps might still be standing! *"TWO IN, THREE IN!"*

His turn, *"FOUR IN!"*, he aimed for the last one, the one that King had pretty well shot up. There wasn't much left of it, but he could make believe! He started firing early, *TATATATATATATAT*, nose up a little, next jeep, *TATATATATAT*, next one *TATATATATAT*, next, until the guns stopped and he nearly scraped the top of the wreckage with the bottom of his plane. That's how the P-51's did it during the War. *"FOUR OFF!" "HAVING FUN, FOUR?" Oh shit! "ROGER, FOUR!"*

Nothing else was said as they joined up for the return flight. As they approached the base, Frits thought about the events of just an hour ago and his heartbeat started up again. *"Would the tower have reported his long and jumpy take-off?"*

"Would # 3 talk?" "I've got to shut him up!"

As they crawled out of their cockpits, he caught up with #3; "Don't say anything about my take-off, o.k.?"

"What happened?"

"I'll tell you later. Don't say anything!"

Nobody was the wiser, except the one for whom it counted... . *he'd NEVER FORGET!*

January 1954 in ARIZONA was marvelous!

They didn't see as much of Skipper and Scottsdale as they would have liked, but they did end up at one of her friend's farm one Saturday, way out in the desert. About ten of them were invited to the "spread", not far from the "LOST DUTCHMAN'S MINE", somewhere beyond CAMELBACK MOUNTAIN. The whole gang got saddled up and a real "COW-POKE" took them way out into the desert, teaching them how to ride "WESTERN" style. And... he took them back. Although navigation should have been second nature to them by then, they would probably have gotten lost in that desert, had it not been for that "COW-POKE!"

A good size steer was being bar-b-que-ed and smelled so appetizingly good, that they could hardly wait for dinner.

The evening in the desert was nice and cool, millions of stars were painted overhead on a huge canvas it seemed, setting the stage for lots of Western songs, lots of beef and lots of beer. They sat around and sang until the wee hours of the morning, alternating between Dutch and American songs.

It was one of the most unforgettable days in the U.S.A!

Then the rains came! It rained, rained, rained as if the ARK would come by there any minute. All flying stopped. The streets on base were just rivers. The guys waded to chow, the F-84 Club or the movies. Pants rolled up, shoes and socks in hand under their ponchos, they sloshed through the knee deep waters to their destinations.

Thank God, it rained warm water!

The first flight after the rains was like a miracle!

The desert looked like Holland: streams, creeks and rivers everywhere! One main difference. Most of the desert streams were real torrents, where in Holland they would have been real placid.

Next day, next miracle! NO MORE RIVERS! Just green, wet riverbeds where the streams were the day before.

One more day and the whole desert was in bloom!

Heaven knows where all those plants came from!

Within the week, the flowers were gone. Just green riverbeds was all there was left and a week later, nothing! Just plain old desert again.

One of those "flowery" mornings, Hob Nob flight went on an early mission. Pre-dawn take-off, up to 14,000 feet and all the way to the GRAND CANYON. The sun just came up over the horizon as they got there and the rising sun created an immense shadowed crack in the plateau. The visibility downward into the canyon was very poor because of the darkness, but the enormity of the canyon was enhanced by the sharp contrast of dark and light!

They made a big sweeping turn while descending to 2,000 feet and because of the "loose battle formation" they were flying, the boys had plenty of time to take picture after imposing picture.

On the way back, the sun behind them, they attacked a "Korean" convoy on an unfamiliar range, way north of their usual practice area.

This was not desert country, lots of trees and the convoy was extremely well hidden. It made for an entirely different mission. Their "success rate" was lower than usual, but all in all, they still shot the "hell" out of those "chinks!"

The nightly war stories in the club never ceased and neither did the beer flow. By mid-January, they were winding down and finishing up. They crated their belongings and shipped them off to New York by rail. The only things they kept with them were clothes for the heat in Arizona and the cold in New York, plus the ordinary necessities.

CHAPTER XVI
GOODBYE AMERICA!

On January 31st they boarded a DC 4 in Phoenix for New York with stopovers in Dallas and Washington D.C.

The excitement about going home was greatly dampened by the awesome sense of loss, leaving all those memories and good times behind them. The BROWNS in Towson, Md., the ROBINSONS in Goldsboro N.C., Capt. ROXS and PIETER in Texas and now SKIPPER ROSS in Arizona. With a pang, he thought of Mary-hyphen-Ann! For a moment he felt weak... . Jo Stafford was singing to his subconscious; *"When You're Near, So Help Me Dear, Chills Run Up My Spine!"*

He did feel an actual chill as he looked down for the last time on all those familiar landmarks, disappearing fast behind them.

It had been so unbelievably great! Such an enormous experience! A one in a million adventure!

Jo Stafford's words kept ringing in his ears; *"Don't You Know , I Love You So, I Wont Be Happy Until You're Mine!"*

He wondered if he would ever see that young effervescent face again.

The boys were unusually quiet. Just five of them were leaving at the same time, some of them had hardly been together since they sailed across on the same ship a year and a half ago.

They had trained at different bases and weren't reunited until LUKE where they met up for the final stage of their training.

Jantje Baetens was the baby of the group. He was one of the few Dutchmen who did not speak decent English when he came over to America. The story was that his buddies coaxed and pulled him through all his classes, but that he was a natural pilot.

He was boyishly good looking with black hair and blue eyes. His rendition of "*SOUTH OF THE BORDER*" could put Frank Sinatra to shame and it was a showstopper, whereever they went. He was without doubt, the most likable fellow in the group. His whole appearance projected honest warmth and everyone took to him immediately.

Karel van der Veen was typically "FRISIAN" in his appearance. Straight blond strawlike hair, blue eyes and very solid features. He had a good sense of humor, but he would sit back and enjoy the antics of his buddies, rather than participate in them and he was not the type to ever initiate anything.

On the way to the airport and throughout the wait, they had been very uncomfortable in their heavy blue uniforms, but once on the plane, they could take off jackets and ties that they hadn't worn since New Years.

As the landscape changed and they watched the snow-capped "Rockies", their moods took a change too. They started to look forward to their five day stay in New York.

On the sixth day, the "VEENDAM" would leave for Holland, but they had no illusions about their return cruise. The boat would be absolutely void of tourists at that time of year and the Atlantic would be chilly and rainy at best. More likely, stormy and wavy!

New York would therefore be the highlight of the trip and the cruise home would be a chance to rest up!

After that: "HOLLAND!", the family, the beer, the friends, the beer, it was something to think about that made their hearts beat faster. In reality, they had all been too busy to ever be homesick, but now that the return trip was "ON", it gave them the shivers.

Frits had been pretty good about writing. Not like in the beginning when they were restricted to the base, but at least "fair!" At "SINTERKLAAS"-time, he had written Mama, "*Take a hundred guilders out of my account and buy some presents!*".

At birthdays he'd instruct Mam: "*Give him or her 50 guil-*

ders!" and he'd write a nice letter to the celebrant!.

In his own mind, he'd been a good son and brother.

Once on the plane, with the reality sinking in that his American tour was over, he got real anxious to get it over with and to get home. All of a sudden he felt a longing for home and his family! He had never known that feeling before.

It all changed in Dallas.

A new flight crew came aboard and he discovered another Mary-A!

She was a typical stewardess. Greeted everybody with that "*YOU'RE VERY IMPORTANT*!" smile and didn't give him the least little hint, whether she liked him or not! Not until he asked her for a "*HIGHBALL*". He was pretty sure that there was no alcohol aboard, but she smiled; "*YES, SIR*!" and reappeared with a tall glass with a brownish liquid, ice cubes, cherry and stirrer.

"HERE YOU ARE,SIR!"

Same standard smile! Not a hint of anything personal behind those blue eyes.

"*THANKS*!" He was pleasantly surprised. This was his first time on a civilian airplane and he was very pleased with himself. He sat back, took a sip,,,,and nearly spit it out!

"*ICE TEA!*" It struck him so funny that he laughed out loud and shook up the passengers around him.

That broke the ice. By the time they landed in Washington D.C. and in the half-empty plane to New York, they set the date.

Carol and one of her co-Stews, Alice, would meet them in the lobby of the Edison Hotel the next evening at 7PM.

By the time they arrived in New York, it was 12:30 AM and too late to do anything but get to the hotel in Manhattan.

A cab ride got them into the city and the Edison Hotel on 54th street, smack in the middle of the city. Another Dutchman was waiting for them, Jaap Heersema. He had washed out of pilot training but had been sent to Alabama for Radar-Training and was going to travel back with them. He was impeccably dressed in a well pressed uniform and starched shirt, right down to his corporal-stripes. All six of them were assigned to a two-bedroom suite with a connecting bathroom. Looked good to them!

New York City was vibrant and alive as they observed from the cab and no way did anybody want to go to bed. They dumped their luggage, got out their raincoats and hit the street. "BROADWAY, BROADWAY!" Doing a shuffle on the sidewalk seemed appropriate. The 45 degree temperature and the icy drizzle felt like Holland and nobody cared. The pubs they hit weren't really much to brag about, but they were in "New York" and that's what counted. They sang a little, laughed a lot and sacked out about 4 AM, sleeping the dreary morning away.

Chinatown lunch, a soggy stroll through "Wall Street" that left them unimpressed and then for a "nickel" on the ferry to Staten Island. Miss Liberty on the way back and that they did find impressive!

Frits and Frits made sure that they were back at the Edison to meet the girls. They were so soaked and chilled that it took a hot shower to revive them. They had also brought some booze to the room in order not to become too dehydrated.

He had a few snorts and went down to the lobby to wait for the girls. The girls were certainly prompt and looked gorgeous in their dresses. Much better than their drab uniforms.

Frits suggested, they'd come up for a drink but the answer was a firm "*NO, THANKS!*" He was totally amazed! He started off defensively; *"THERE'S ONLY THE SIX OF US, HAVING A DRINK!"*

"FIVE!", Carol corrected him. At first he thought she lost a marble or two but then it dawned on him that she was unaware of Jaap's existence. She had served just FIVE of them on the airplane.

The answer remained; *"WE'D RATHER NOT!"*

So the girls waited in the overstuffed chairs in the lobby while Frits went back up to get Frits.

Well, Frits was already fairly sloshed, so Henk volunteered to replace him and neither Frits objected.

The girls might have liked dinner and a show, but with three more days in New York and eight more on the boat ahead of them, that didn't fit into the boy's budget at all so they ended up in a very cozy East Side Pub. At least the girls knew their way around town. The Pub turned out to be a pleasant discovery. It seemed so far removed from Broadway and Times Square and it seemed so much like a neighborhood tavern, that they felt immensely at home. Instead of being surrounded by tourists, they were surrounded by "locals", New Yorkers!

Other pubs along the East Side in the 50's and the 60's proved to be very nice as well.

Whatever the reason, the evening didn't go all that well.

Maybe the guys talked too much about themselves and bored the girls with their "flight" talk? Who knows? Maybe the girls heard enough of that amongst their colleagues. Maybe the boys should have bought them dinner. Anyway, about eleven, the girls got up, shook hands and thanked them for a "great time!" Bullshit!

A bar patron who overheard some of their conversation thought the guys were speaking German and suggested "Yorkville" to them.

He explained that was "Germantown" on 86th street and they decided to explore that area. That turned out to be a winner. The "HOFBRAUHAUS" reverberated with the bangs and booms of the hoompah band, pounding out the "beerbarrel polka!" They could handle that!

Most of the crowd were New Yorkers of European descent and the rest were happy tourists. Henk wasn't into Polka's and Waltzes that much, but to Frits it was balsam to his ears. The Waltz was the first dance he learned when he was only six and after he had danced one waltz, girls between the ages of 20 and 80 started to ask him to dance! He could do that all night and never tire!

After 2 AM they crossed the street with a bunch of partygoers from Queens to the "GEIGER" and they changed from dancing to singing. It was too bad the girls had left so early, they would have enjoyed it, he was sure!

After their third successful performance of "*Lustig Ist Das Zigeuner Leben, Fariah*!", they were shoved out of the door and there was nothing left to do but go to bed. It was 4 AM. On the way to the subway station, they happily sang: *",It's Quarter To Three,There's No-one In The Place, But Just You An' Me. So Set 'Em Up Joe!"*

There was no question about it, Henk wasted his money on singing lessons in his youth! Or maybe it was just the beer!

Before they knew it, it was their last night in New York and simultaneously, the end of their money.

They were expecting to receive some kind of a check upon boarding, but for now all they could do was pool their money and buy a bottle of Gilbey's Gin and some lousy mixers. They packed and drank, they joked and drank, sang songs and drank and as the evening wore on, their singing and laughing got louder and their jokes more rambunctious. Truth of the matter was, they got rather sloshed!

A few of them decided to take one last stroll down Broadway and get some fresh air. That was their most brillant idea of the whole America trip. The nippy air and the brisk walk had a sobering effect on them and besides that, the gin was gone by the time they got back..

Jantje was lying in the bathtub, zonked out. In one room, a fully clothed body lay across a bed, out like a light and in the other room the laughter was so loud, it was a miracle that "security" hadn't been summoned to put a stop to it. Karel was in hysterics. He tried to retrieve the last drop from a glass by sticking his tongue in it as far as it would go! The guys who had been out walking couldn't quite figure out what was so funny, but the laughter was contagious.

Jan de Ruiter, tears running down his cheeks and doubling up with laughter was trying to explain what was so hilarious, but he couldn't get the words out!

It was time for bed!

Jaap and Frits retreated to the other room, took the shoes off the body on the bed, (the other Frits) and covered him up.

The laughter in the other room continued for a while and Frits heard a shower running before he dozed off. He remembered thinking: "Somebody's sobering up!"

At 5 O'clock, either a siren outside or a bladder problem woke him and with all the neon lights outside he didn't even have to flip a light on. He finished his thing in the bathroom and was just about to retreat to bed when he noticed Jantje Beatens still laid out in the bathtub, but now the the tub was full of water!

"HOTDAMN!" It wasn't cold in the room, but he shivered anyway. He leaped the two steps to the tub and reached in to pull the plug with one hand while he felt Jantjes forehead with the other. He was not dead! He was cold, but he was breathing!

Frits ran back into the room and shook Jaap: *"COME HERE!"*

He flipped on lights and dashed back to the bathroom, Jaap behind him. *"Pick him up! Let's put 'em on the floor!"*

Each grabbed an end and lifted the dripping body onto the bathroom floor.

"Put a blanket on the floor in the bedroom and come back to help me!"

With feverish haste, he started unbuttoning Jantje's shirt, threw it aside, his T-shirt next! Jaap was back and started to take off Jantje's shoes and socks while Frits cursed the "STUPID BRITISH BUTTONS" on Jantje's fly. Jaap pulled the pants off, the shorts came next and the two of them picked up the naked body and carried it to the bedroom where they wrapped him up in the blanket.

"Get some towels! If he throws up, catch it with a towel! I'll be right back!"

He shot into his shirt and pants, slipped on his shoes without socks, grabbed the wet clothes and dashed out to the elevator. In the basement he turned down the corridor to the cleaners. The Chinese lady smiled knowingly when he said; "*ACCIDENT! HURRY*!" and he was gone!"

They only had one set of "blues" with them, the rest had been shipped in their advanced luggage, so Jantje had to wear these pants!

Back in the room, Henk was massaging one leg and Jaap the other.

" Good show! Did he wake up?"

"Nope, what happened?" Jaap wants to know.

"I don't know, ask Henk! He was there!"

"I don't remember. Karel turned on the water and thought it was hysterical and so did I, but I thought he turned it off!

"Somebody did, because the tub was only half full which was a good thing because the poor bastard might have drowned, goddamnit!

"The water was cold as a bitch when I drained it, he must be stiff as a board! Is he warming some?"

Frits reached under the blanket for his arm.

"Look at that! Waterproof watch! It's still ticking!"

He slipped the watch off and took Jantje's pulse… .. "72" he announced! *"THAT SOUNDS NORMAL*!" Jaap observed.

"Well, put a pillow under his head and let him sleep. I'll shave and shower now, so I'll be out of your way. We'll have to eat by 8 anyway!"

At nine they left. They half-carried Jantje. They had to go by a warehouse on the west-side to retrieve their luggage that had come in by train, before moving on to the ship in Hoboken.

Jantje at least carried his own hangover and so did some of the others. Filing out of the hotel, they did not make a very glorious exodus.

At the docks they ran into an unpleasant surprise! They

had to pay $48 storage fees.

There they were....*SOPHISTICATED JET-FIGHTER PILOTS.... CONQUERORS OF THE BLUE YONDER.....COSMOPOLITAN INTERNATIONAL TRAVELERS*......but did not have $48 between them.

They had hung on to enough money to manage cab fare to Hoboken, but just barely!

They tried to get the management of the storage place to bill the Dutch Consulate....N.G..... they tried to call the Consul.... not available...

"We're gonna miss our boat....this stuff was all pre-paid freight...and......"

"You're right! You paid for the freight but not for 6 days of storage!"

"Goddamn....what can we do? We can't leave without it!"

A meek voice asked; *"Would you pay me back?"* It was Jaap.

"You have money???? How much have you got?"

"$300."

"You have $300??"

"Off course we'll pay you back! Pay the man!" What a solution!!! *"Off to the seven seas!!"*

On board each had a check waiting for $42.50, which was worthless until the bars opened, three miles out to sea.

It was only 10 AM and the boat wouldn't shove off till noon, so they put Jantje in a bunk without breaking anything and Frits crawled in the next bunk with the instructions; *"Wake me at noon, I don't wanna miss Miss Liberty!"*

He just dozed off when Flieger shook him. *"Listen to this! "I have the passenger list! The ship holds 750 passengers, right?"*

"If you say so!"

"There are only 300 booked, o.k.?"

"O.K.! I'm not surprised this time of year!"

Flieger was grinning from ear to ear,,,*"Here comes the zinger!!!!*

"""""ta ta"""""

"What is it?"

"Ta ta" Of the 300 passengers, 54 are men" "Ta ta"

"What?"

He sat up so fast, he banged his head against the upper bunk!

"Are you shittin' me?"

"Ta ta!" Flieger threw him the passenger list.

"Let's get out of here! No, no, not you Jantje. We'll wake you later!"

He had never gotten into his clothes so fast in his life!

"How do you figure all those women all by themselves? They're all "Mrs." Nearly all of them are married!"

"Wives heading for Germany to join their husbands!"

"You've got to be kidding! Hey, wait Frits! We've got to play it cool! First... .. don't pick on any newly weds! They're still true! We've got to find out which ones haven't seen their husbands for a long time cause they're hot to trot! So don't go falling in love immediately, interrogate 'em first!"

"FritsForrer, did anyone ever tell you, you talk too much? Let's go look 'em over!"

There were very few "BON VOYAGE" wishers on the grubby docks of Hoboken and consequently, there were very few passengers on deck. In the lounges, finger sandwiches, pastries, coffee and tea were being served and....complimentary drinks!!!! *"AH!!"*

The bars could not sell anything until they were out at sea, but they certainly could serve them free!

Now that was more like it!

The waiter suggested a *"KOPSTOOTJE"* for them because of the cold weather. That was a new one on them!

A "KOPSTOOTJE?"

It literally meant; *"A SHOT IN THE HEAD"* , but it sounded good. It turned out to be a shot of "Jenever" with a beer chaser.

Nobody suggested that it was too early in the day to drink and besides, how could they argue with the Captain's orders?

The "BOKMA" Jenever was the first they had tasted since leaving Holland and it warmed them up from the inside out.

It really gave their intestines a warm glow that spread to all their extremities, especially their head! To top it all, the "HEINEKEN" draft was so smooth and tasty, that they had to sit back and savor it slowly.

"My God, I forgot how good that stuff tastes! Remind me to take that nap this afternoon, but for now " another kopstootje please, before I catch a cold!" They nearly forgot their mission: "*Check out the girls*!"

"250 of them?" "How many single?"

"Who the hell wants to chase single girls? They're too damn complicated!" Imitating a woman's voice; *"I'm not that kind of girl, I don't kiss on the first date!"* He was on a roll. *"I have to know you better and if we do, you know?, you won't respect me anymore!"*

Everyone broke up laughing!

"No,no, what we need on this trip is a sexstarved female who wants to screw day and night, so they'll have to roll us down the gangplank because we're too weak to walk!" "Agreed?" "Let's find them!"

Obviously, nobody wore a button; *"KISS ME, IM SEX-STARVED!"*

The "hunting" would make it more intriguing.

When the boat started moving they attempted to arouse Jantje. To no avail. They gave up trying and went on deck to wave goodbye to Manhattan, Miss Liberty, Staten Island, Brooklyn and finally, fading in the mist, Coney Island.

It was a very touching moment with a lot of sadness gripping their hearts. The other passengers must have felt the same way, because the mood at lunch was very subdued, not festive at all.

After a few more "pilsjes" Frits went to sleep, looking forward to a long evening and night!

It turned out to be an interesting cruise indeed. Fantastic food, good music, pleasant company and Jaap to borrow money from whenever they went broke. Being outnumbered "five to one" gave the boys an unusual amount of attention and although nobody surfaced with a "SEXSTARVED" button, enough "hanky-panky went on to keep the male passengers and the ship's officers smiling.

Dancing nearly every dance that was played kept down the weight and sweated out the alcohol, so their endurance proved to be outstanding.

One morning, at about ten, someone shook him and said: *"We're in Holland, wake up, we're in HOLLAND!"*; all he could think of was *"Turn this boat around, lets go back!"*

Not only were they about to leave their temporary paradise, the gray, foggy, dreary scenery outside was such a depressing view, that he wished he were back in Arizona.

The boat however, continued into Rotterdam!

All but Frits had relatives meeting them at the dock, so he was left to help many of their female friends with tickets and connections and finally at 4 PM, he headed home.

He hadn't written that he was coming on that day, as a matter of fact, he had lead them to believe he wouldn't be home until the end of February, rather than in the middle.

He was looking forward to surprising them!.

CHAPTER XVII
BACK HOME.

Once alone in the train compartment, he started to get real excited about seeing everyone and as the train loaded and unloaded people in DELFT, THE HAGUE, LEIDEN, and HAARLEM, he was so absorbed in his thoughts that he didn't notice the glances people threw in his direction. He was a strange sight to see.

Where all the other passengers were as pale as milk bottles, he was deeply tanned. He had no insignias on his uniform indicating any kind of rank, but did have a strip on the top of each sleeve: "NETHERLANDS" and two "WINGS" on his chest. On the left his orange and gold "VINK" and the American silver "WING" on his right.

In his mind he toyed with pictures of his brothers and sisters, what they should look like now a year and a half later.

A pang of guilt hit him, thinking he should have called from Rotterdam, making sure everyone would be home.

There was little he could do about it now and it was dark when he got off the train in ALKMAAR and lugged his stuff over to the bus.

"Oh, hi Frits! You're back!" The bus driver was one of his buddy's older brother.

"Sure am!" shaking hands.

"Let me help you with that luggage! Sit up here!"

He lowered the jump seat that no one was supposed to sit on once the bus was in motion.

"Tell me, how were things in America and where have you been and do all the girls in America look like moviestars and drive convertibles? Did you go to Hollywood while you were there and what are you flying now?" And on and on and on!

Frits couldn't get a word in edgewise, but nodded and waved at other passengers that he recognized or who recognized him, or both.

Being the son of the bar owner in town and being on the soccer team made him well known in the small community of St. Pancras.

The trip barely took 15 minutes and there he stood in front of the familiar sign; "CAFE RUSTWAT". The door flung open as if he were expected after all. His luggage was picked up and loving arms hugged him, warm lips kissed him and he was hustled inside.

The welcome was fabulous.

"Why didn't you take a cab? Why didn't you call?"

"Have a beer, brother!"

"My God, you look good! You've gained weight, you look more solid!"

"My, do you have a tan! You make all of us look as pale as snowmen!"

"Let the man sit down, he must be tired!"

"Are you hungry? Clasien, fix him a plate!"

"Have another beer!"

He couldn't get a word in edgewise here either, but he thought to himself; *"God, it's good to see them all again! They've gotten to be so big!!"*

His baby sister was nearly 8, Hedwig was going on 17 and with her boyish haircut, she looked cute as a button. Brother Joop had just turned 24 and looked like a gentleman. Paul was just 19 and would soon be drafted. Clasien was 20 already! My God, nobody stood still, except Mama. Mama looked younger. The hard years through Papa's illness were far behind her and she glowed with health and pride.

Some friends and neighbors dropped in and Frits found out that they had indeed expected him. Ever since 4 in the afternoon actually! What had happened: Wim Oosterbeek, the boy that had washed out way back in Bryan, had called at 4 and asked to speak with Frits. When he was told; *"FRITS IS IN AMERICA!"*, he answered; *"I DON'T THINK SO, HE'S SUPPOSED TO HAVE DOCKED IN ROTTERDAM AT 11 THIS MORNING!"*

Well, that set some tongues wagging and Mama decided; *"I BET YOU THAT WIM IS RIGHT! HE'S GONNA TRY TO SURPRISE US!"*

It worked out fine. The suitcase was opened and presents dug out. There was something for everyone, although Hedwig believed that she got the most precious gift of all: a silver and turquoise belt, made by Arizona Navahoes. After all, she was his favorite!

It was impossible to tell all the stories, listen to all the happenings from back home and get all the questions answered before the night ran out on them.

Somewhere along the line Hedwig asked; *"You still don't smoke?"*

"Nope!, I've got enough bad habits!"

Paul and Hedwig had both taken up smoking since he'd left and both of them put their cigarettes out and he didn't see them light another one throughout the night. At least he had a positive influence in some ways!

They sat around the big table for hours, basking in the happy glow that existed that evening until Mama shoo-ed everybody to bed at 1 o'clock; *"It's going to be an early day for everybody*!" and she was right!

Everybody had to go to work and Engelien had to go to school. She should have been in bed long ago! A few more hugs and everyone meandered up the stairs to the great big attic and Frits turned to his old room that he shared with Paul and crawled in bed, exhausted but feeling wonderful.

It was good to be home! He had not thought about it very often when he was gone but as he crawled between the sheets he realized "THERE'S A LOT OF LOVE AND WARMTH UNDER THESE OLD RAFTERS!" and his eyes turned moist before he fell asleep. He felt he was a very lucky man!

Over loud protests, he got up with the rest in the morning; "You can sleep. You don't have to get up!"

He wouldn't have missed it for the world. Nothing had changed. It was still very much old world style. Mama, as always, was the first one downstairs, dressed for the day. Hair done, skirt and blouse, shoes and stockings, ready to go! The only time he had ever seen his mother in a nightgown and "peignoir" was during air raids or when Papa was sick. As a matter of fact, only on the night that Papa died!

That was it! Otherwise that lady was presentably dressed at all times.

The banter and chatter at breakfast was invigorating to him. It was not a community breakfast like in the old days because of the different work schedules and departure times, but that kinda made it more fun. He must have sat at that breakfast table for two hours enjoying the troops making it through breakfast and getting out by bike or bus. Finally only Mama and Hedwig were left, Hedwig because she didn't have to leave till 10 in order to be on time for her lunch duty in the "STATIONVIEW" restaurant.

Predictably, Mama said;" *Shouldn't you shave and get dressed?"*

Nothing had changed!

Nope! Mam would always be the same, which was O.K.! She was bred tough and after that: marriage, six children, business, depression, German occupation, war and a sickly husband had made her tougher yet.

Before she even opened the bar doors for business, she sat Frits down and accounted for his money.

The 30 guilders a month, that he had agreed to pay his Mam when the government subsidy stopped, all the "fifties" and "hundreds" that he had authorized for birthdays etc. had reduced his total from a potential of f 3,900.-to f 1,400.-

Boy! It was all there and it was all correct, although in the back of his mind he expected about f 2,000.—

Well, it was all accounted for, so when his Mam handed him his bankbook, he told her *"Thanks for taking care of things for me and from now on, I'll just send you thirty guilders a month or do you need more?"*

"No, no! Everybody is pitching in and I'm getting caught up on all my bills. As a matter of fact, things are starting to look rosy for the first time in my life!"

"It's about time, isn't it?"

She shook her head and sighed a sigh soooo deep, it must have come from her midriff. *"IT SURE IS!"*

It was nearly ten years ago that the Germans took away their business and not till 1947, three years later, did they get off the welfare rolls when they bought this money losing place.

It had been rough. At 48, widowed, grey hair showing through seemed to prove that fact!

Frits took the bus and the bankbook. He hadn't answered her question; *"What are you going to do with the money?"*

Just; *"I'm going to look around town. I need to buy some shoes for this kind o' weather."*

In OUDDORP he got off the bus at the old church and in a steady drizzle, looked for his Father's grave. At first he couldn't find it because he wasn't looking for a headstone and was about to give up when he spotted: "HERMANUS JOHANNUS FORRER".

The headstone had been added of late. He tried to pray, he tried to feel emotional but it all meant nothing to him. He made up his mind he would never come back there.

In the bar next to the church, he tinkered with the billiard balls, sipping a beer while waiting for the bus.

The owner was tickled to see him and called out to his wife; *"Look who's back!"*

It was good to be back home!

In town, in ALKMAAR, he went to a used car dealer and looked at what was available in his price bracket. The dealer, whom he had known for years, convinced him that a little Ford Anglia was his best choice. Frits would rather have had a V.W., but they cost three times as much with the same kinda mileage. Much to his surprise, financing was no problem at all, so while the dealer got things ready, Frits went to the bank, withdrew f 1000.-, gave 800 guilders to the dealer and drove off.

He was the only person in the entire family to own a car, uncles and aunts included. He didn't have a license to drive it yet, but that didn't really bother him much. At age 22, he held a valid U.S.A.F. and Netherlands Military Pilots license, so what was a little driver's license?

He had a car and that was all that mattered. The hell with being poor! He was going to live a little! He put on his new shoes and went for lunch in Hedwig's place. She just went GA-GA over him and smiling from ear to ear, she introduced him to her boss so loudly, that the whole restaurant knew him instantly.

Shooting some billiards at Jan de Groot's bar, he caught up on some of the latest gossip.

Jan de Groot's place was where they would shoot billiards

on Sunday mornings when Mama believed them to be in church across the street. He found out who died, who got married and what good looking girls got onto the scene.

A visit later on to the local department store, VROOM & DREESMAN, confirmed that some of his old friends were still around and still good looking. It made him marvel at the fact that there were so many good looking girls in town. With a car and a few guilders in his pocket, he ought to make out like a bandit!

By mid afternoon, he decided to go home and take a nap like he had done on the ship each day. It helped him persevere through the nights. It didn't work out that way though.

A couple of old buddies were already shooting billiards, hanging around and waiting to see him!

Most of them worked the land in the surrounding polders and didn't have much to do until after the spring thaw. The grocer's son was kinda his own boss, so he was there and one of the girls from a family of three daughters that he had dated was there too!

It made for a real old fashioned homecoming.

There was no end to the questions nor to the beer flow and when he brought out hundreds of pictures that he hadn't sent home yet, plus the hundreds that he had sent home, his friends really received a grand tour of the United States from the Atlantic to the Pacific. They treated him as if a big time war hero had come home from years in battle, hung with decorations.

Thank God, Dutch tradition prevailed and just about everyone went home for supper, so the family could have somewhat of a meal together. Hedwig was not home from work yet and somebody had to stay behind the bar and take care of the stragglers.

It must have been 7 o'clock before someone asked; *"Whose car out front?"*

"Oh, mine!"

"Yours? You've got to be kidding! Can we sit in it?"

Even Mama couldn't keep the gang at the table this time, so she grabbed a shawl and joined the enthusiastic gang out front. Frits ended up driving seven people around the block in his four-seater while supper was getting cold.

After a few days and a lot of beer, things slowed down some as the novelty wore off. People were getting used to having him back.

On Saturday night he got talked into going dancing in Beverwijk with Joop and some of his cronies. Normally, the guys would have gone there on their bikes, about an hours ride. Now they went by car. Joop asked if he could drive, because after all, he had a license and Frits didn't

Well, that was the first time and the last, as far as Frits was concerned! He was too erratic. No wonder they hadn't wanted him for a pilot!

The dance club "'t Vosje", "the Little Fox", was very nice and as predicted, loaded with women.

He stood out like a sore thumb. All Dutchmen wore a tie at all times, certainly when going out. The chilly weather called for heavy tweeds and woolens, but all Frits had on were tropical things and his "Perry Como" shirt and suntan added to the exotic impression he made. He just gloated over the attention he was getting.

One problem surfaced immediately. Being in a car with four guys was like being back in America, where boys and girls arrive in cars together and have to go home again, boys with boys and girls with girls. Not very many good chances to make out!

There was no way he was going to be able to make out with "AALTJE", VOS' daughter, because she had to help clean the joint and Frits would have to drive the other guys home. He decided right then and there; "From here on in he would prowl alone!"

Meanwhile they danced their shoes off, laughed a lot and made a lot of new acquaintances. It was a fun night and long after the music stopped he promised Aaltje, he'd be back in the morning. After all, it was only 30 minutes by car! Thank God for a small country!

In the morning, when he left for "church", he made 'tVosje in 25 minutes. Because the owner and Frits' parents belonged to the same Restaurant Association, "HO-RE-CA", the "VOS" himself knew his folks very well and went out of his way to make Frits feel at home and his 21 year old daughter certainly helped to make him want to come back for more.

After playing some billiards with the locals, he sat at the bar chatting with Aaltje, her Dad and Corrie, a waitress who had worked the dance the night before and stayed over. He was asked to stay over for Sunday dinner, but he couldn't do that to his Mam. She expected him back for Sunday dinner. That was a lifelong tradition.

He set out a little later than he wanted to and was asked to drop off Corrie two towns down the road. At first he was a little annoyed because he really wanted to make tracks, but once in the car, Corrie came on so strong, that they were trying to smooch while he was driving and within ten minutes he turned off onto a dirt road. They couldn't wait to get down on a grassy spot and they tried to bang the living hell out of one another in the wildest sexual interlude he could remember!

The whole thing didn't last but 10 minutes and he had to stop at a gas station to wash up, so he wouldn't walk in smelling of raw sex!

Is that the way it's gonna be in Holland? Is that the way it was when I left?" He asked himself: *I may never go back to America!"*

A few more days of R & R and the vacation was over. He had to report to "VOLKEL" AFB in the same good old southern province of BRABANT, but this time in the extreme North East corner, rather than in the extreme South West at WOENSDRECHT where he started his flying career.

He looked forward to strapping himself to an F-84.

He envisioned himself tearing up the skies over Holland within the next day or two.

He couldn't have been more wrong!

After getting housed in familiar German built brick quarters, they reported to the 313 Squadron, the training squadron.

A gentle, soft spoken Indonesian Pilot/Navigator, Major Benjamin was in charge and a fanatical Navigator, Major Sjoerdsma headed up the ground school. The first thing he did was send Frits out of the room to have some Sergeant's stripes sewn on.

Frits couldn't remember what he did with the stripes they'd given him in Bryan, Texas, 7 months ago.

The next thing the Major did was point out the window at a T-33 and said; *"Gentlemen,,,that is an airplane! In about 3 more weeks you may get to touch one, but then only with an instructor!"*

They had earned their Wings on that machine a long time ago and after that flew many, many Gunnery missions on it and that miserable S.O.B. told them they couldn't even touch it!

Putting more salt in the wound he went on; *"First you're*

going to learn how to fly!"

That did it! Their American classmates were already fighting in actual combat over Korea and this "hornrimmed" bastard insinuated that they couldn't even fly!

Some of them would have resigned from the Royal Netherlands Air Force right then and there, if only they could have.

That was depressing!!!!! The major however was unperturbed.

He started his navigation, computer-protractor and meteo classes and continued his damned demeaning attitude.

There was something that mollified the boys to the point that they nearly forgot their resentment. He taught them things that they were completely unaware of.

The Dutch weather for one. There were only 100 days a year when they could expect clear skies. That was different than 300+ clear days that they had gotten accustomed to.

Now they learned how to cope with the other 265 days. The weather in Holland was such that most of the time only elements of two planes would climb out through 10,000, 15,000 and sometimes 25,000' of clouds. All radar controlled with exact speeds, headings and altitudes until they were "on top" where they could do some flying!

Navigation was different too.

No "RADIO-RANGES". Instead; "QUEENY GEORGE HOW", Q.G.H., with which radio controllers brought them down through the soup until G.C.A, Ground Controlled Approach, picked them up at 2,000' to bring them in.

Terminology became different. "Angels 5" meant 5,000 feet!

Spotting a "BOGIE", unknown or unfriendly aircraft, evoked cries of: "*Tally-ho, there goes the bastard*!" The British way.

Navigation over the flatlands of Western Europe was quite different too, but it had a positive aspect. All the many waterways and lakes were like street signs!

They could tell where they were as clearly as if the sign had said: "56th and Broadway!"

The study assignments were not easy and their torturer gave them plenty of it.

The boys had to admit when it was all over, *THEY HAD LEARNED AN AWFUL LOT*!

Socially, things were fine.

They caught up to a lot of old familiar faces in the N.C.O.

mess and met some new ones. The town VOLKEL, was just a bunch of farmhouses it seemed, but right outside the gate was a little joint called "MOEKE PEREBOOM", or "LIL' MOTHER PEARTREE".

The food was good and cheap, they could walk to their beds if need be and the beer was plentiful. The place loaded up with Airforce people of all ranks and functions and there was no end to the war stories, past or present, in the air or in bed!

One night they hit the Southern town of EINDHOVEN, the home of "PHILLIPS" radio and bulbs. A dynamite metropolis! The market square was, as in all European towns, lined with bars, clubs and restaurants. In one of them, the "POORT VAN CLEEF", the pilots from the A.F.Bases EINDHOVEN and VOLKEL congregated.

Over the bar was a string of hooks with beer mugs dangling from them. The pilots had individualized them by scratching their name on the bottom and so everyone drank from his own mug.

There was a jazzy dance club next door, the "BENELUX" and after downing a few beers in those places and some others, Frits stated on the way back to Volkel; *"When we're finished with this school, I'm gonna transfer to Eindhoven!"*

That was his kinda town! The southern, nearly Belgian, relaxed, easy going attitude of the population was his kinda cup of tea! Of course he had no control over his upcoming assignment, but he did put in for Eindhoven anyway, knowing that half of them would end up there eventually.

Holland had two tactical bases for ground support; "VOLKEL"and "EINDHOVEN."

There were three "AIR DEFENSE BASES" with GLOSTER-METEORS as their primary fighters; SOESTERBERG, known to Americans as "CAMP AMSTERDAM", smack in the center of Holland, "TWENTHE", along the German border and "LEEUWARDEN", up north near the North Sea.

"GILZE-RYEN" was the main Dutch training base where Dutchmen earned their Wings on T-6's.

Primary schools with Tiger Moths were: HILVERSUM, near SOESTERBERG, WOENSDRECHT where Frits got started and YPENBURG near THE HAGUE. YPENBURG also had Piper Cubs, but the main Piper-Cub force was at DEELEN, north of ARNHEM, where the British landed in September of '44, as part of the "MARKET-GARDEN" invasion and were beat all to hell.

DEELEN also had the first "Bell" helicopters.

The NAVY had one airbase, just north of THE HAGUE, called VALKENBURG.

Considering that Holland was only 200 miles wide from the North Sea to the German border and only 300 miles long from the sea to Belgium, it had great airbase coverage.

Scattered here and there were some civilian fields, SCHIPHOL at AMSTERDAM of course the biggest.

Besides all of those, there were numerous auxiliary fields, actually slabs of concrete laid down by the Germans during the War.

What all that meant was: eyes open and roaming at all times when flying V.F.R., (Visual Flight Rules) and strict adherence to I.F.R. flight rules, radar and radio controls when flying on instruments.

In the U.S. they had gained instrument experience under the hood in T-6's, T-28's and T-33's as well as a lot of link-trainer simulated time.

Their actual flying in the clouds was NIL. In Holland, that would change fast. Nine out of ten flights involved some "actual" instrument time, mostly climbing out or letting back down through the clouds.

Instrument training therefore became one of the highest priorities. Instrument flying under the hood was a tense business to say the least, scary even on occasion and downright terrorizing when vertigo played tricks on the senses. When a pilot's senses tell him that he's upside down, but his instruments tell him he's right side up and straight and level, he's got the damnedest time believing his instruments. BUT....HE'S GOT TO!

Like a race driver in a banked curve, a pilot is as comfortable in a coordinated turn as if he were sitting at a desk. Rolling smoothly into a turn creates no sensation whatsoever. Rolling out of that turn suddenly starts the fluids in the equilibrium moving, creating the sensation of a roll in the opposite direction, while the instruments show perfectly level flight.

In the clouds, as in a fog, there are no references, so a pilot cannot tell by looking out of his cockpit, whether he's standing on his tail going straight up or standing on his nose, going straight down.

BELIEVING IN HIS INSTRUMENTS.. is his first MUST!

The instruments might lag a little, but they don't lie!

The worst thing that could happen in a cloud was for a wingman to lose his leader. A wingman, staring at his leaders tip tank light had no way of knowing whether he was turning or level, the only thing he could tell by airspeed and engine noises was whether he was going fast or slow. That was it. Just the wing light lined up with his leaders helmet and no more. No way of knowing whether he was level, climbing or descending, just hanging in there. He also didn't need to know. All he had to do is make sure he stayed in formation, a few feet apart and not to lose his leader. If for some reason he were to lose sight of him, he was to throttle back immediately and get some space between them to avoid a collision and immediately get on the instruments. He'd have to interpret the plane's attitude and interpret it right or else he might make the wrong moves and end up in an "unusual" attitude from which he might not recover.

"Unusual Attitudes" were therefore practiced endlessly.

"*Close your eyes*!" The instructor then would make a bunch of turns, climbs, dives and whatever else he could think of, leaving the pilot so confused about his attitude in the sky that it resembled the game; "PIN THE TAIL ON THE DONKEY!"

"*You've got it!*"

Eyes open immediately… .. situation… . speed dropping … . artificial horizon in a high bank… altimeter climbing…

gyro compass turning fast… *STEEP CLIMBING TURN*!!

Nose down first!! Power!!! Level wings!!! Adjust nose to level flight!!! At cruise speed, power back to normal!!!

A next time... Altimeter unwinding... speed increasing fast… . compass turning… . artificial horizon shows plane way down left... Solution!!!! Pull back power while leveling wings!!!!! Pull up to level flight!!!! When level, power to normal cruise!!!

If a pilot spent his time leveling his wings in a steep climb, he might stall out and in the clouds, that would be deadly. *NOSE DOWN FIRST*!!!!

When in a turning dive, if he pulled on the stick first, he would only tighten up his turn and he might lose it all together. So... *LEVEL WINGS FIRST*!!!

A pilots reaction could only be one way... *THE RIGHT WAY*!!!

Reaction the wrong way could prove to be deadly and it often was!

In order to make it more interesting, the instructors would turn off the power to the instruments so all the electrically and vacuum driven instruments wouldn't function. The only ones that would function would be the natural ones, the ones that depended on air pressures. Altimeter, vertical speed, airspeed and of course the magnetic compass.

That was hairy, to say the least.

Under the hood, in back of the T-33, they practiced and practiced until it became second nature.

Formation flying in the T-33 was next. Then: landings, take-offs and aerobatics and finally...SOLO!!

After having flown "solo" all alone for three months, they solo-ed all over again. First on T-33's and then on F-84's.

Some new aspect was added: the guys got to fly together in the same plane for the first time. They had never had anyone in the backseat as a passenger, other than their instructor.

Strangely enough, there was some apprehension!

Over a beer a guy would whisper; *"I dont want to ride in his backseat, I dont trust that guy!"*

Frits was flabbergasted! Even if the guy seemed nutty on the ground, he was still a licensed pilot and deserved to be respected as such! He'd volunteer: *"I'll ride with him*!" and that satisfied everyone without any bruised feelings.

On some of those flights they would send four buddies in two T-33's on a low level navigation flight all around the country, with the guy in back playing navigator. He'd do all the plotting and figuring and the guy up front would maintain course and altitude. It was as much fun as a bunch of guys racing around a parking lot in "go-carts".

Dog fighting in F-84's was another new aspect. An instructor would try to get on some new pilots tail and while pressing the transmitter button, he'd go : *Tatatatata", you're dead 495*!"

Of course, in their own minds the new pilots were "HOT SHOTS" and they would do their damnedest to keep from getting "shot down", better yet, get on their instructor's tail and go: *"Tatatata, you'r dead blue leader!"*

Another new fun sport was "trail formation" where all four planes would fly behind and below the other, spaced out a few airplane lengths. It was like "follow the leader" or "Simon says". Whatever the leader did, so did the other three behind him with just a split second delayed reaction. If #-1 would start a "bar-relroll", within a second #-2 would roll, then 3 and of course, #-4 working his tail off trying to keep up!

#-4 remained the hardest position and because his throttle setting would go from "100%" to "full-closed and dive brakes", he would burn more fuel than the others. So it was up to #-4 to keep his flight leader informed: *"#-4, TIPTANKS DRY!"* so the flight leader could make sure they'd be back on the runway before #-4 flamed-out.

The amount of flying time varied very much with altitudes and performances and the guys had to learn to figure out exactly how much fuel they had to burn.

An F-84 with tip tanks loaded could climb to 40,000 feet, cruise at "MACH 0.83" and stay aloft for $2^1/_2$ hours. With the same amount of fuel in simulated "COMBAT", chasing one another all over the sky, from ground level to 20,000 feet, they'd burn that fuel in 50 minutes.

They learned an awful lot in that 313 training squadron and when it was all over, they were very thankful they did.

They knew they were better pilots bcause of it !

They received their permanent assignments; Jan Mante, Jantje Beatens, Karel van den Veen and Frits Forrer to:

314 SQUADRON, EINDHOVEN!

WHOOPEE!!!

CHAPTER XVIII
EINDHOVEN

Nearly everybody had bought a car. Frits Flieger, who stayed at Volkel, had bought a "SKODA", a Czech car, Jantje Baetens had a "Hansa-Borgward", Frits had his Ford Anglia and so the guys teamed up in their wide variety of makes and models to transfer over to Eindhoven. Karel had not bought a car, he was thinking "saving and marriage", so he rode with Frits in the little Anglia.

The first thing that was different at Eindhoven was the quarters. All the N.C.O.'s were in a camp off base with the pilots in the last building down the street, just past the bar and the mess.

The first thing that was the same was the German built fortress like buildings with the thick brick walls. This time, no bunk beds, all individual beds.

The place was called "CAMP BEATRIX" after the Crown Princess of the Netherlands.

After dumping their stuff, they headed for the base which was a good ten minutes away.

One of the first buildings they encountered on the base was the 314 "OPS" (operations) building and crewroom. In the OPS office they were met by a 1st lieutenant named SOESMAN.

He looked half Indonesian, was very underweight and had the damnedest habit: he stuttered on the vowels instead of the consonants. He introduced himself as Lieu-eu-eutenant

Soe-oe-oesman! He didn't call Frits; *"F-F-Forrer, but Fo-o-orrer,* not *B-B-Baetens,* but *Bae-ae-aetens*! He was the oldest First Lieutenant they had ever met. He had to be at least 45 and must have been a Non-Com Pilot for a long time. Like so many of the Senior Officers in the Royal Netherlands Air Force, he had probably served in the "KNIL", the original Royal Indonesian Army Airforce.

During World War II, most Dutch Pilots fled to England and flew with the British against Germany. Some had learned to fly in England and all of them were allowed to wear the R.A.F. wings along with their Dutch "VINK", just like our boys were entitled to wear the USAF wings.

The Dutch Pilots that flew in Indonesia got out to Australia and flew with the "AUSSIES" against the Japs. Some were even attached to units of the U.S.A.A.C.

The tragic part was that little recruiting of pilots took place for about 10 years, from about '39 to '49, so there was a distinct GENERATION GAP between the old "prop-pilots" generation that ran the Air Force and the young "Jet-Jockies" of the last few years.

It was a little bit like "Grandma and Grandpa" being forced to bring up their grandchildren! It was not the healthiest of situations.

Soesman took them into the Squadron Commander's office and did the introductions. "CAPTAIN CHARLES POUBLON", small, stocky, infectious smile, slightly balding up front and covered with dark curly hair everywhere else, stood up behind his desk and extended a hand and made them feel at home immediately. That was a relief after meeting Soesman, who definitely made them feel ill at ease.

Capt "POOP" welcomed them to the "best Squadron in Holland" and told them they were expected to be "TOP-NOTCH" to belong and they in turn assured him, that was just "fine" by them.

Poublon was obviously one of the many "Indonesian born" Dutchman, his accent was so thick, you could cut it with a knife.

He took the boys around back to the crew room and ,as usual, they met some new faces and some old familiar ones. Red Verdonk was one, Pierre Gilliam had been ahead of them in Kinston and Henk Benninga, already the silver "dot" of a warrant officer.

Frits became a member of "BLUE FLIGHT" with as it's leader MATJE WILLEMS, a short, baby faced 2nd lieutenant. He didn't look a day over 19 and as it turned out, he hated that. He was a serious young man but like Napoleon, always felt the need to flaunt his importance in order to offset his youth and size. Henk Benninga and Red Verdonk were also in Blue Flight and the Squadron call-sign was "FANCIER", so little Frits ended up as "Fancier Blue Four!" He could live with that!

The squadron was very young, formed in 1952 and Frits' picture on the "Pilots board" was only # 25!

The plane that was assigned to him was the "AT-21", an F-84G, the latest and the most modern of the F-84 line! He was happy!

Their job was pretty much an 8 to 5 job. They were on their own as to the time to get up, shave or shower, to eat breakfast or not eat breakfast. At 8AM, they were expected to be on base, in their seats in the "briefing room" for their morning "METEO" briefing. This was new to them. In the U.S. the weather had generally always been "flyable" and there was hardly any need to sit through an extensive weather briefing.

In Holland though it was quite different. The weather over Europe was generally so stinko, it was highly necessary that the pilots knew what to expect and how to deal with it. Most days, take-off ceiling and visibility limits were just marginal and they might end up landing at alternate bases if their own socked in.

Next were safety tips by the "safety officer", another old lieutenant. Aircraft identification was another must. They were shown dozens of silhouettes of aircraft of all sizes and makes. The projections showed those planes from all possible angles and they were expected to identify them. Being only one hour from the "IRON CURTAIN", it was imperative that they could tell the difference between a MIG-15 and an F-84 when they were looking down it's tailpipe, although they were damn near identical from that angle. From that angle it could also be a French MYSTERE for that matter! With all the different types that the Russians produced, MIGS, YAKS, SUKHOI'S etc., the Swedish LANSEN, all the British planes; the GLOSTER METEOR, CANBERRA, HAWKER HUNTERS, the American F-86's, B-29's B-25's, B-52's, B-57's and on and on and on! There didn't seem to be an end to it! With the jet-age speeds, it was most important that they'd identify a plane properly at first sighting because it was only a matter of seconds before they'd be in range of one another's guns. Coming at each other at a combined speed of 1,000 MPH, it sure wouldn't take long!

While they were sitting in the briefing room nursing hangovers and identifying silhouettes, the ground crews were getting the planes ready, so by the time the boys meandered over to the crew room for coffee, 4 or 5 planes would already be "serviceable" and Soe-oe-oesman's voice would already sound over the intercom; *"BLUE-UE-UE FLIGHT, REPO-O-ORT TO O-OPS!"* or ; *"WI-I-ILLEMS A-AND FO-O-ORRER, REPO-ORT TO O-OPS!"*

During the course of an average flying day, 15 out of the 25 planes would be serviceable, but sometimes only 8 or nine! Of course, every pilot wanted to fly as much as possible, but for the Dutch Pilots there was another incentive! They received 8 guilders per hour additional for each "JETHOUR", so they could double their salary by flying 30 hours a month.

The Flight leaders, Willems, Brosky, Nieuwlands and De Jong flew a lot more than the newcomers of course. The money aspect of flying kinda made it a wholesale scramble to kiss Soesman's ass because he was the man in control. Frits hated the thought. If someone wanted his friendship or respect, they had to earn it! Mere rank only proved that they had stuck it out longer, not because of what they had accomplished. In many cases the upper ranking Officers had simply not wanted to face that cruel world out there: civilian life!

Soe-oe-oesman was alright maybe, but Frits had no respect for him from the start and it never improved. Too bad! He would have been better off being Soesman's buddy.

Matje Willems proved to be a stickler for detail. His preflight briefings and de-briefings were long and precise and he demanded perfection. While flying, that was alright. On the ground with tedious paperwork, he managed to bore Frits to tears and in reverse, he could get really ticked off at his newest addition to his flight for not knowing a procedure or policy by heart. He was a sharp pilot though! He went strictly by the books instead of "by the seat of his pants."

Right off the bat they started practicing battle-turns and simulated Air to Air patterns until blue flight performed them to perfection. This was all in anticipation of the upcoming Air to Air gunnery that they would practice shortly over the North Sea, north of Leeuwarden.

As in the U.S., high-level-battle was flown in two "elements" of two; "lead" with #2, and #3 with #4 on his wing. The two elements would fly just about abreast, a quarter of a mile apart, their wingmen on the outside about 5 airplane lenghts off their wing and behind. The whole idea was; if the lead element went after a "bogie", the second element would follow far enough behind to also be able to get in a pass at the bogie and keep other potential bogies off the leaders ass. #4 also protected his element leader from potential enemy fire but who would protect #4 was not quite clear. Frits felt a little like a sacrificial lamb.

While flying straight and level, battle formation was a cinch and a pleasure. Making turns was more complicated. If the lead element made a 90 degree turn away from the element and that second element tried to stay on the same side of the leader, there wasn't enough power in the world to allow him to keep up in that much wider turn. So, as soon as the leader rolled into his bank, the element would roll into an even steeper bank, cross below and behind the leader and roll out abreast of them again, but on the other side, a quarter of a mile out. The wingmen meanwhile had slipped to the other side of their element leaders, again to be on the outside.

So, if done perfectly, a flight that was seen from left to right, flew in this order #2,lead, #3 and 4, before the turn would look like #4,3,lead and 2 after 90 degrees If the leader continued the turn and thus started to turn into the second element, the reverse became true. If nothing was done, the element would make the shorter inside turn and end up way ahead of the leader, so he slipped ahead and below the leader and rolled out perfectly abreast again.

Matje made them practice and practice until near perfection. Henk, being the element leader was getting to be real good at it and Frits just had to make sure that he stayed in his position of Henk's wing.

Air to Air gunnery practice did not go so well. Matje would send out Henk on a straight course at 200 MPH and then the three of them would make simulated gunnery passes at him. It was the same basic pattern as in LUKE, but Frits did not fly the required tight patterns to Matje's satisfaction.

Frits convinced himself that the mid-air collision back at LUKE was not on his mind. No, no factor at all, but it probably was. He didn't tell Matje about the collision so they kept on practicing and practicing!

Evening time in Eindhoven was more pleasant than it had been in Volkel. The bar at Kamp Beatrix was more cosy and since the "new" pilots didn't seem to have a chip on their shoulders, the other non-coms readily accepted them and made it fun for them to be there.

Practically next door was the Catholic clubhouse; "THE TJI-TJAK", (Indonesian salamander) run by Chaplain Major FRITSIE. He was of World war II vintage.

He loved his "JENEVER" gin and he had a broad understanding and a very open mind when it came to Military Human Frailties, like "women". He could tell dirty stories with the best of them, his beer and booze was cheap and he tried to keep "his" boys close to home in his "Tji-tjak" where he could watch them. Everyone was encouraged to bring their girls to his place and dance to his records, play ping-pong or just sit around and talk. He provided a quiet section where a guy could write, read or study.

Town itself was about a 15 minute ride by car and Eindhoven literally had hundreds of bars. Just in the market area there must have been 50 of them.

The "POORT VAN CLEEF" was their starting point usually and the owner, short, chubby "DIKKE WILLEM" did his damnedest to attract the pilots. The only son of the longtime owner of the place, he had been spoiled enough to have taken private flying lessons in his teens, which in Holland must have cost him a fortune.

The atmosphere in the club was certainly conducive to partying because of the decor and the beer mugs. "WIM" gave Frits a mug on one of his first visits and with a ballpoint pen he scratched his name on the bottom of it along with some of his philosophy that Aristotle would have been proud of; *"THERE'S JUST ONE LIFE; THAT'S A GOOD LIFE!"* Wim lacquered the bottom with a clear varnish and voila; a personalized mug that would withstand a hundred years of drinking and rinsing.

Most week-ends they ended up going home, not necessarily because of loyalty to their mothers, but because they had girlfriends back home.

Frits still ran around with "Aaltje" most week-ends but could not get over the amount of beautiful and willing women there were in Holland. One week-end he ended up in "Zwolle", ancient city on the river "IJSSEL" as the guest of Red Verdonk. Red had been a star water polo player and a "free-style" swimming champion, so he was a hometown hero with all the fringe benefits that come with that honor; lots of girls! Besides that, the lean and ruggedly handsome redhead had a sense of humor that wouldn't quit. When Frits met his redhaired dad and younger brother Mario, he knew that it ran in the family! They had a blast all week-end.

Jantje Baetens lived in Zaandam, just 20 Kilometers south of Frits' home and Jantje had love problems. In itself it was not an unusual problem in Holland in those days; "PREJUDICE!" There was still a lot of animosity between religions and the major political parties in Holland are Protestant and Catholic.

Jantje had maintained correspondence with his girlfriend during the year and a half that he was gone and apparently true love bloomed between them! The problem became, she was Catholic and he was Lutheran. That wouldn't work! His folks were adamant. "HE WAS NOT TO SEE HER AND CERTAINLY NOT MARRY HER!" He was only 21 and certainly had plenty of time, but,,,, they were in love! Jantje was broken up over the whole contraversy because he loved his parents very dearly too!

One week-end, a long week-end, he called Frits at home and asked him if he could pick him up at his girlfriends house in Amsterdam on his way down. He ran his "Hansa-Borgward" underneath the rear of a truck that stopped at a railroad crossing and his hood was as flat as a pancake but was being fixed.

Once a month, the R.N.A.F. had a "long week-end" from Friday at 4 PM till Monday morning. "DUE BACK TIME" was: "BY THE EARLIEST MEANS OF PUBLIC TRANSPORTATION AFTER 8 AM."

A local guy would be in by 8:30 if a bus left from his frontdoor at 8:05. Frits could catch the 8:15 in St, Pancras, catch a train in Alkmaar, arrive in Eindhoven at 11:40 and the bus would get him to Kamp Beatrix by 12:30, so HIS due time was 1 pm.

Because of this arrangement, people came in by dribs and drabs all morning and by 1PM rollcall, 95 % would be "present or accounted for".

On this particular Monday morning, Frits had breakfast with the rest of the family, pulled out at 8 and was at Marjan's door at 8:30, right on schedule. He honked the horn and Marjan appeared in the 3rd floor window; *"He'll be right there!"*

There was no place to park in that city that was built for bicycles and streetcars.

Jan was down in minutes, dumped his baggage in the backseat and started talking about his problems.

He had had a real drag down fight with his parents and had spent the week-end with Marjan and her parents. That should have made it fun, but it didn't. He was really torn up about the whole thing. During the entire two hour trip to the base he talked about his love for Marjan and how much he cared for his folks. While changing into their uniforms and on the short trip to the crew room, he kept bending Frits' ear about the situation and Frits felt the best thing he could do was give him a shoulder to cry on, so he patiently lent him his.

At the crew room, with hardly a soul present, the intercom blasted: *"BLACKFLIGHT TO OPS!"* Four of them left, including Jantje and Frits started to fix himself a cup of coffee.

"FO-ORRER,O-OPS!" Soesman was assigning planes as they became available from maintainance and another one must have been just called in.

Soes handed Frits the clipboard to the AT-25 plus his mission assignment. On the board Frits noticed that "HIS" plane the AT-21, had been assigned to Jantje, so he told Soesman;

"I can catch up with blackflight and trade planes with Baetens, right?"

"Why-y that shi-i-t?"

"Because the "21" is my plane and the "25" is Baetens, that's why!"

"I-i do-ont gi-ive a damn who-o-ose pla-ane it is! Go-o and fly-yy!"

"Bastard!"

He caught the little V.W. van that took him to the dispersal area (no flight lines in Holland) and waved at Jantje as he was busy starting up the AT-21.

The planes each had their own concrete dispersal area, surrounded by earthen walls, to protect the planes from bomb damage unless they received a direct hit.

The weather was sunny with just a few little patches of white and the assignment he'd been given would keep him up there for at least an hour and a half. One and a half hours of absolute freedom!

Black Flight took off ahead of him and after shoving the throttle to 98% power, he released the brakes and he was off!

The sky was his oyster!

He climbed to altitude, homed in on the 1400 foot radio mast in the center of Holland, requested an outbound course, like he was supposed to and decided to go say "Hello" to Mama!

That was tricky for more than one reason.

He had no orders to be anywhere near the North Western coast of Holland, they were not allowed below 1000 feet over land and if someone reported him, it wouldn't be hard for the authorities to figure out who was buzzing Saint Pancras. It might spell trouble, so… .. he went!

He came down upon the house with full power and dive brakes down in order to cause even more of a rumble. As he passed over the house he pulled up steeply in a left hand turn and looking over his shoulder he could tell Mama already had customers. There was a truck out front and a few bicycles and right then, people were coming out of the front door.

He pulled back on the power, popped the dive brakes again and slowed her down to 250 MPH. This time he came around and passed in front of the house at 500', paralleling the street. He rocked his wings as he passed in review! Power full forward, dive brakes up and into a left climbing turn. He rolled over to his right, rolling onto his back and pulled through, back into the direction of home, paralleling the street again. Still upside down, he reached 500 MPH rolled out and roared in front of the house, then pulling up nearly vertical, while turning slow rolls toward heaven. After 3 vertical rolls and at 7000' he headed away, as fast as the plane would fly. That had been enough. No point in overdoing it and getting caught!

Back at 15.000' and over the same radio station, he requested a Q.G.H. letdown and the radio operator brought him home.

When he reported over I.P., the tower instructed him to circle the base and come in from the opposite direction. The runway had been changed. For a moment it puzzled him, because when the tower gave him barometric pressure and wind direction, it hadn't changed from an hour and a half ago, but he followed instructions and landed from the opposite direction from which he had taken off. He taxied back to his dispersal and he was still busy throwing "OFF" switches when a green military V.W.Bug with his squadron Commander "POOP" in the passenger seat.

"Oh shit! Already?"

Someone reported him and he got caught, pants down!

Poublon was up the ladder before Frits could unbuckle!

"*OH, SHIT*!" was all he could think.

Poop asked him, the moment their eyes were at the same level; *"How well do you know Jan Baeten's parents?"*

"His parents? I met them once briefly, why?"

"We need you to ride with the Chaplain and me to talk to his parents! He just clobbered in!"

"He what?"

"Ya, man. Right in front of the runway! Come on! By the way, who would know his parents better than you?"

"Karel van den Veen sir, they've been buddies for years!"

"Is he in yet?"

"I dont know. Should be!"

He climbed in back of the bug with the Chaplain.

"Captain Poublon, you should also notify his girlfriend in Amsterdam. They had wedding plans even!"

"Goddamnit! Sorry Reverend!"

"That's quite alright. I know this is upsetting!"

"Forrer, do you have the name and address and phone number of this girl?"

"Yes sir! It's in my car."

The bug pulled up to OPS.

"Let me get it, I'll be right in!"

Passing by Soes he hollered; *"I'll sign in in a sec, o.k.?"*

In Poop's office, POUBLON looked deflated.

"Forrer, they've located van denVeen. I've sent for him, tell me about the girl!"

Frits related briefly Jantje's story about Marjan and Poop said four more times; *"Goddamnit, sorry Reverend!"*

The Reverend just nodded, didn't say a word.

He was the Protestant Chaplain, whom Frits had never met.

Just when Frits asked; "*What happened*?", Karel came in, white as a ghost.

"Goddamn, man, I'm sorry!" Poop was shaking hands with Karel, *"I hear you were real buddies!"*

"Yes, sir! All through flight school. Nearly two years!"

His voice was barely audible and everyone was still standing.

"Sit down, man!" Poop's Indonesian accent seemed more pronounced.

"I've got to sign in my plane, sir! Excuse me!"

Frits started out the door while Poop's hollering; *"Get back in here!"*

Frits signed in, turned in the clipboard and after knocking, stepped back into Poop's office. Karel looked like he was ready to cry.

"Actually man, (Poop pronounced it mahn) *Both of you should come, but it's such a small damn car! Karel, do you feel up to it?"*

Karel just nodded!

"O.K., MAN, YOU GO!" He pushed the intercom; *"Soes, give Forrer a plane!"* He didn't seem to expect an answer, Frits saluted and left.

Soes handed him a clipboard; *"You-ou have ha-a-alf tanks, be ba-a-ack in an hour!"*

"Yes, sir!" Two flights in a row, what a luxury!

The bus driver only knew that the plane dove into the ground in front of the runway. The crewchief had more info; *"He rolled upside down from maybe 300 feet and clobbered in. She didn't burn or anything but she broke into about a dozen pieces!"*

"How come?" Frits wanted to know, *"Why did he roll over?"*

"Aileron-booster trouble!" He had heard, *"That's why they came back so soon!"*

It didn't make much sense, but the moment his wheels were up, he swung to the right of the runway, keeping his nose down to pick up speed and at the end of the runway he put the plane on it's left wing so he could get a clear view.

A dozen vehicles surrounded the wreck, ambulances, towtrucks and firetrucks, about a mile off the end of the runway.

"JANTJE BAETENS, BE WELL, WHEREVER YOU ARE!" That was all he could think of, he couldn't think of another thing to say!

He climbed out until he was out of view of the tower and radar and set course for Germany.

The Germans had lost the war, had nothing to say about anything and "low-level" was allowed over Germany. The moment he crossed the border, he put the plane on the deck. A glance at his chart confirmed that there were no high tension wires anywhere near, so he flew between the treetops wherever possible. This is what he loved! If only he were on a mission with loaded guns and a target to go after, he'd sneak up to the target, pull up and dive down, shooting the living shit out of it! That had to be the way to live! Too soon, he had to head back or risk running short of fuel.

In the crew room he heard the story! It sounded too stupid to be true!

While at altitude, flying in formation, Jantje complained about a heavy wing. The flight leader sent the element on it's way by it self and joined up on Jantjes wing.

First he had him "run-up" the "aileron boost", the power steering. When the aileron-boost was at a minimum, it gave the pilot a three to one ratio . In other words, the muscle power that the pilot supplied was multiplied by 3. It made the plane harder to handle but was great on instruments and cross country flights, it made the plane steadier.

For "dog-fights", "gunnery passes" and "aerobatics", the ratio could be run up all the way to 13 to 1. Just a touch of a feather would then roll the aircraft.

So when Jantje "ran-it-up", it did make it easier for him to hold the aircraft level.

The Base Safety Officer got up in the tower and got involved. He had Jan run the booster up and down and had him violently rock the plane, just in case one of the tip tanks wasn't feeding. It might get the fuel flow going, if one tip tank fed and the other stayed full. Nothing changed.

The Safety Officer seemed convinced that it was booster trouble and Jantje believed him. HE TURNED OFF THE BOOSTER!

He should have trusted his own judgement, but that was Jantje for you! An officer always knows better! Horseshit!

So, he brought her home. His flight leader on his wing, they came in for a straight-in landing. With gear, dive brakes and flaps down, they decreased speed on final and to his leaders horror, about a mile out, Jan slowly rolled over and clobbered into the ground right in front of his eyes. He was killed instantly!

All he would have had to do was push the tip tank release button, the tip tanks would have fallen off and Jantje and the plane would have been saved. Yes, one of the tip tanks wasn't feeding.

Or, if he had kept the booster on and on a high, 13-1 ratio, he might have been able to hold the plane and landed! What a stupid way to go!

Frits was adamant; *"The pilot is in ultimate command and makes the final decisions, not the man in the tower and not the flightleader either!"*

He was right about that!

"I would have said: "Screw you!" and I would have dropped the tiptanks! No good? Fiddle with the booster! Still no good? Goodby plane! Hello parachute!"

They talked about it for hours and hours and they all agreed; *He should have dropped his tanks, even if it proved that nothing was wrong with them*!

Too bad! Jantje was a natural pilot, a real good one!

Everyone agreed, they would have used their own judgement, rather than relying on ground advise.

"Goddamn Soesman!," In the bar later on over a beer, " I asked him to trade planes with Jantje, I had the 25 and he had the 21 and we could easily have switched, but no, not Soe-oe-oesman! Jantje would have been alive right now and I would have had to face the problem with the tiptanks. I would have ejected the sonsabitches!" "That lousy crum Soe-oe-oesman"

Nothing brought Jantje back! Four days later they were in

The funeral of Jan Baettens

Zaandam for a full military funeral.

"What a stupid thing to do", Frits was saying to himself.

Some twenty pilots and officers marched behind the caisson and a complete 100 man peleton of soldiers with rifles behind them.

"Those parents must be torturing themselves!", He was thinking, *"Look at all those healthy young men! Why mine? Why my Jantje? Why not one of them?"* Frits felt it would have hurt less if there hadn't been any military at all!

But,,,, they were there! Frits was one of the pall-bearers. That didn't bother him yet. Seeing the casket go down didn't hit him that hard either. Neither did the 6 gun salute. The eulogy was close to getting to him, but he controlled himself.

Then came, in his opinion, the most stupid part of the whole ceremony: passing by the grave and dumping a shovel full of dirt onto the casket, making a spooky, hollow sound that got him annoyed.

But when Marjan walked up, after Mr. and Mrs. Baetens had done their thing, and instead of taking the shovel, she threw her bouquet down on the casket as if to say; *"Damn my fucking fate*!"

Tears just started rolling! Standing at attention, he couldn't do a damn thing about it either.

He was grateful to see Poublon salute rather than throw dirt and when it was his turn, that's what he did. It just seemed more appropriate.

GOODBY JOHNNY, I'LL SING "DOWN MEXICO WAY" FOR YOU, MANY, MANY TIMES. MANY TIMES!"

CHAPTER XIX
314 IN DENMARK

When the new contingent came in, it was like an explosion!

There they came: SHORTY HORSTEN, all 6'4" of him. "IKE" EIJKELENKAMP, KEESJE BOS, JOOP JONKERS, HARM APPELO and KARELTJE TEN HOVE!. The noise level in the crew room went up by 50 decibels! The Air Force staff had selected a cross section of the best CLOWNS and some of the heartiest PARTYGOERS in the business!

When one of the "old time" officers attempted to act seriously, just one look at "Shorty's" crooked grin made him forget all about his honest intentions!

Now, going into town meant that the one squadron could fill up an entire bar! Poublon would sometimes come into the crew room just to hear Shorty explain a situation. His arm would swing around like windmills, his tall body would literally double up with laughter: "*HE HE HE HE HE HE HE*!", at his own nonsense!

His laugh was so infectious, there was no way anyone could keep a straight face when he laughed.

One of the most positive things about them was: they were all good pilots, every one of them! A crop like that was rare

They weren't in the squadron very long when the 314 and 315 Squadrons went on maneuvers with all available planes.

For two weeks they would rough it on an auxiliary base,

"BEEK", in the southern part of Limburg.

BEEK was only 72 miles to the South East, but the logistics involved, moving all the mechanical personnel and equipment, kitchen personnel and equipment, plus the medical staff, operating tables, bunks etc.! It was a chore! Not for the pilots though, they just flew over with some spare clothes and toiletries.

Tents would be pitched in an apple orchard and everything would be run out of tents: Ops, briefings, mess, clinic, everything!

Lieutenant-Colonel Hansen, the assistant base commander, ran the flight briefing personally and assigned the formations within the formation and concluded: *"The runway is short, but definitely long enough! No sweat! The runway lies on top of a hill and when you're still a few hundred feet up, you see it in it's entire length, but when you're close to touchdown, all you can see is the part that runs uphill. Don't panic!!!!! Don't slam on the brakes!!!!!! There's plenty of space left once you're on top of the hill!!! So relax and we'll be alright!!!*

It was like war! The guys were issued 7mm Browning pistols in shoulder and armpit holsters and heavy desert boots.

During the 314 meeting "Poop" told the whole squadron; *"Our squadron is going to be different! We're gonna grow mustaches, all of us for the duration of the exercise!"*

It was going to be fun!

Thirty two planes took off in pairs and joined up in boxes of four. Then four boxes of 4-plane-boxes made up the 16 planes from each squadron, 32 in all from 314 and 315.

Col. Hansen flew in the lead plane and when they reached their destination, they started a wide circle around the field, while flights of 4 peeled off to start their approaches to the field, shifting into echelon formations for landing.

Hansen first. As he flared out for landing, he saw the runway rushing toward him, he slammed the plane down at the very end, slammed on the brakes, blew a tire, ran off the runway to his left and broke off a main wheel as he came to a stop in the grass. He looked like a crippled duck!

His #2 man landed in fine shape behind him, but when he saw Hansen run off to the left instead of rolling straight ahead, he panicked and ran off the runway to his right, cracked a wheel in the soft grass and looked like another crippled duck!

#3 landed O.K., but must have really stomped on his brakes

and locked them, so he blew two tires and stopped, obstructing the runway! The rest of the planes were diverted to "HAHN" A.F.B., an American base outside of Cologne, Germany.

So far, they were losing the war!

Later that day, the planes returned to BEEK and landed without a hitch! Hansen didn't show his face much.

Living in tents with only their legs for transportation made the whole affair somewhat of a boy scout outing.

Half of the time the weather was minimal and since there was no radar on base, the mobile units had to provide guidance in and out of the base and that proved to be minimal too.

Some days, they just did nothing.

On one such socked-in day, some unexpected entertainment came from the farm house, adjacent to the apple orchard.

A mare was being "studded".

Before long, some 300 airmen lined the barbed wire fence, encouraging the young lovers.

They were big horses, Belgian plow horses and it was assumed that some good money was being paid for the "studding" service.

The encouragement from the military ranks was not helpful.

"GO BABY, YOU CAN DO IT! GO FOR IT!"

"HEY SOESMAN, IS THAT AS BIG AS YOURS?"

"WANT ME TO COME OUT AND SHOW HIM HOW?"

"WHY DONT YOU TICKLE HIS BALLS?"

"LET'S GIVE HIM A HAND!"

The horse got nervous and didn't do anything, just stood there on his hind legs, his front hoofs draped down the mare's back, prick inserted and .....nothing!

Well, the stud was lead away and reappeared after a little while. A hand was run underneath the mare's behind and then stuck under the studs nose and, as by magic, that long dingeling started to reappear, nearly two foot of it! The ranks cheered and applauded! *"THAT'S WHAT YOU NEED, HARRY! LET HARRY SMELL THAT HAND*!" The farmer guided the penis back in there by hand and got a lot of cheers. *"SEE, SOESMAN, THAT'S HOW IT'S DONE! NOW YOU KNOW !* And on and on and on!

It was one of the most hysterical events of the whole exercise!"

On one of the socked in days, when trucks returned to the base in Eindhoven for supplies, Red and Frits rode along for

Squadron sex ed

the "overnighter". Hitting the bars in a flight suit and a pistol in a shoulder holster seemed like newer version of some old "Wild West" stories, but they had a blast, drank too much beer and did not hurt anyone in the process. Only their heads didn't feel so good in the morning.

On the weekend, trucks were provided for transportation to the beautiful and ancient city of VALKENBURG and like the city was overrun by Romans and Germans at one time, it was now overrun by Dutch Airmen. They were out for fun and they made a good job of it. Singing and laughter could be heard coming out of every bar and restaurant and when the trucks loaded up again at 5 PM, there was not a sober face in sight! Drivers excepted.

During their second week in tents, Poop confronted Frits: *"Goddamnit man, you're supposed to grow a mustache like everyone else!"*

"I am, sir, I am! I haven't shaved under my nose at all!"

After a close inspection, Poop exclaimed: *"You're hopeless,man! You have no hair on your upper lip! You'll never make a good pilot!"*

The exercise proved that a move like that could be made,

On maneuvers

but that was all. Back to Eindhoven! This time without crashes!

All of it was a preparatory exercise for their move to Leeuwarden, way up north.

They were going to be up there for two weeks for Air-To-Air, so a lot of guys wanted to take their cars up. That left the rest of them to fly up the planes.

"LEEUWARDEN" was an Air Defense base with Gloster-Meteors, the earliest British Jet Fighters. The Meteors were twin engine jobs, that could out climb, outturn and definitely out dive an F-84. The dive brakes of that bird came out UP and DOWN out of each wing and when the pilot put the plane on it's nose, he damn near had to give it throttle to make it go down.

Since the F-84 was designed for ground support, rather than for Air-to-Air, tackling with a Meteor in a dogfight could prove dangerous to an F-84's pilot's health. Only excellent airmanship could lead to a victory in the sky! Getting on a Meteors tail and pushing the mike button; "*TATATATATATATATA*" was the ultimate thrill of an F-84 hunter.

The runway at Leeuwarden was shorter than Eindhoven, so landing was more precarious, but not as difficult as at Beek.

The F-84's occupied a separate section of the base, not far

Box formation ahead

from the main hangers, with barely more than a crew room to operate out of, but it was cozy. It was more intimate, nearly romantic and certainly adventurous. It was more like camping out!

Poublon started to take an interest in Frits' low scores. He didn't understand it. The main difference between Arizona and the North Sea Islands was the altitude. In Arizona, the pattern was usually from 12,000' and up. Over the North Sea it was from 2,000' and up. Otherwise everything was pretty much the same. Except, it only took minutes to get over the target area, so they had more time over the target. Flights of two, instead of four made it less hectic over the target and easier to concentrate on the business at hand: "shooting holes in a flag!"

Here's where Poop came in: *"Forrer, man, I can't believe that you can't shoot better than 10 or 12 %. I wanna see for myself, man!"*

And he did! He stayed right on Frits' ass throughout the flight, which was remarkable all by itself. He had his own pattern to fly, his own shooting to do and yet found time to find Frits somewhere in the pattern and give him pointers on positions and altitudes; *"UP THERE!" Two, you're too far out! Two, reverse now!, Two, pull off steeper!, Two, you fire too late! Two, close it in! Two, you're too low, get up there!* It went on like that all through the flight. But it helped!

Poop would shoot between 35 and 50% just about all the time and was just about the TOP GUN in the squadron. The other most consistent shooter was Lt. Hans de Jong, the man with the looks of an Italian lover.

On Frits' part, improvement was nearly immediate! He was like a kid in class who'd suddenly see the light in a math problem and completely grasped a new concept and erased all former problems. All of a sudden, like relaxing on the stick in instrument flying long ago, he had it! His patterns became more perfect, his passes were right up to snuff when it came to distance, speed and coordination and the scores started to follow suit. 28%, 30%, 25%, no more 10 and 12%

Then a fluke! He was shooting yellow that day. Everything felt right. He had excellent control. He only pulled the trigger, when he was "dead on" target and then only for a second.

One of two guns jammed after a few passes and he continued with just one gun.

When the flag had been dropped back at the field and all the spent rounds and all the holes had been counted, it caused Poop to exclaim: "*IMPOSSIBLE*!" He ordered a recount and personally walked over to the target drop area to count the yellow-rimmed holes. *NO DOUBT ABOUT IT! A NEW R.N.A.F. RECORD!*

A whopping 83%! Of course Soesman had to knock it; "*IT WAS O-ONLY BECAU-AUSE THE GU-U-UN HAD JAMMED*!" But it was a record just the same! The previous record was 63%, so there was a fair chance that that record would stand for a while.

His scores didn't hit 50% and up anymore there after but he was up there with the best of them, 30's and 40's all the time!

Because of the fact that the married or serious guys coundn't go home every night like back in Eindhoven, there was more of a general party atmosphere in the squadron and the highlight was on Saturday. The rest of the base was off for the weekend and the F-84's had the place to themselves.

A couple of bets were made that sounded crazy as hell. But boy, were they fun! First, Poop won a case of beer for the squadron by diving down at the base at high speed and pulling straight up, making 7 vertical rolls, which was supposed to be impossible in the overweight F-84, heavy as a tank and sadly under powered.

Then, Nol van den Mortel, a Captain from the England era, made a bet with a visiting Belgian Lieutenant. They would fly head-on over the runway at precisely 1000′ and the first one to chicken out and break off would lose a case of beer. Rules were set; exactly 400 MPH, exactly 1000′, dead center over the runway and break-off to the right. They went off!

Ten minutes later, there they came! They were heading dead-on with an 800 MPH approach speed and they seemed to wanna collide over the middle of the field!!! It took only seconds, but just prior to an apparent collision, the Belgian pulled up and right and Nol kept going straight out as if he had a fixation of sorts.

He did! When he was back on the ground amongst his jubilant backers, he explained; *"Well, once I was set perfectly for the runway, I lowered my seat and went on instruments. I never even heard or saw the guy!"*

When flying ceased at 3 PM, the planes were being serviced and the beer was brought out to the flight line where the party lasted till dinnertime.

Everybody from Private to Captain had a ball and it set the stage for more parties in the clubs and bars of Leeuwarden.

The whole gang of Sergeant-Pilots invaded the market place of Leeuwarden and livened up a few joints. They were all in civvies, but somehow the whole world knew they were Air Force guys!

Wonder why?

Leeuwarden, the Capitol of the Province of Friesland, was fiercely proud of it's main products:

"DAIRY and "BEEF!"

Friesland has more lakes and waterways than any other part of Holland and sailboats are docked by the thousands all over the province.

The connecting lakes also provide the setting for the "ELF STEDEN TOCHT", an annual skating event through eleven towns.

Because of it's waterways, no invaders had penetrated Friesland over the years, no Celts, no Huns, no Romans, no Franks, nobody ever conquered Friesland. So, the Friesians are a very pure race. They're very blond, nearly white haired, blue eyed and maintain their own language and their own national anthem.

Every sign, indicating the town or the village is displayed in two languages; Dutch and Friesian. e.a., Leeuwarden = "LJOUWERT".

Their FRISIAN COWS are world renowned as excellent milkers and are the pride of Friesland.

That's why it's understandable that in the middle of the market square, there's a giant statue of a cow with an imposing udder.

It should also be more than understandable that more than just a few people were raving mad on Sunday morning!

There she was in all her glory, the IDOL, the COW, done up beautifully in the colors of the 314 Squadron, right down to a bright red udder.

Why the local police and the M.P.'s came to the base and tried to blame it on the poor pilots was less understandable but the guys had to clean up the cow anyway! OR ELSE!

The pride of the Officers Club was a picture, greatly enlarged, of "PRINS BERNHARD" in the backseat of a Gloster-Meteor. The Prince did have a valid Pilots License but because of his eyesight, he couldn't fly solo, therefore, he'd come out to Leeuwarden to fly the backseat with somebody.

For some reason, the unsportsmanlike Air Defense Pilots sent M.P.'s all the way to Eindhoven to retrieve that picture from the 314 Crew room. No sense of humor!

During that time a few "314 Squadron Songs" evolved.

One of them, a simple incoherent conglomerate of songs that ended with: "LEVE DE KONINGIN, HOY!" (long live the Queen, HOY!")

Frits wrote 314 lyrics to the tune of; *"Into The Air, Junior Birdman!"*, but the simpler ditty won out in popularity and was adopted. Frits really thought that his was greater, but repeated performances didn't change anybody's mind.

Until that time, Sergeant Pilots slated for Officers rank, followed a course at the K.M.A., the Koninklijke Militaire Academy".

The Academy is located in the beautiful medieval city of BREDA, about an hours drive from Eindhoven.The K.M.A. was run by the Army and they didn't have a sense of humor either. At the old fort, Holland's version of West Point, old canons were posted behind even older walls and moats and when one of the canons ended up guarding the 314 Squadron crew room at Eindhoven, the Army did not send an undercover team to retrieve it, no, they sent M.P.'s. What chickenshit! They even threw around terms such as "THEFT" and "COURTMARTIAL"! What a gutless Army!

The 322 Squadron of SOESTERBERG had more of a sense of humor, but was also better prepared. Their mascot, a PARROT, was turned loose in the hangar at night and the stupid bird wouldn't come down from the rafters at some strangers command. Result: no parrot for 314!

Every so often, exchanges would take place. Danes would come to Holland, the Dutch would go to Denmark. The French would exchange on that same kinda basis. And so.... Blue Flight went to Denmark for two weeks!

The Danes had had serious problems. They had F-84's clobber in at the rate of ONE a week! It was blamed on drinking.

They took tough measures. No more "ÖL", their light beer during lunch. Cots were placed in the crew rooms. A pilot who felt hung over in the morning could sleep it off in the crew room. No repercussons. Also; NO MORE FLIGHTS OF MORE THAN TWO AIRCRAFT TOGETHER!

RIDICULOUS? Sure was, but those were the orders in Denmark. So when Matje, Red, Henk and Frits arrived in a tight BOX formation over "KARUP A.F.B." on JUTLAND, the Danish mainland, they put on a little show.

First a pass in the box formation, then a slow climbing barrel roll, ending in an "echelon" formation diving down toward the base, passing over the base at such a low level that the tower operators were looking down on them. At the end of the runway a one-second pitch-up for downwind leg, spacing themselves for landing. ONE, leftside of the runway, TWO on the right, THREE on the left again with FOUR to his right. They were joined up again in tight taxi formation before turning off the runway, THUNDERBIRD style.

It was fun showing off, rubbing it in!

Flying with the Danes was fun and they did a lot of it. When they flew with them, a sole Dutch F-84 would be the #3 man out there, joined up to a Danish element of two.

They really didn't do much of anything in the air because the Danes were barely allowed to do aerobatics, but one thing was exciting! The Danes had to protect their side of the "IRON CURTAIN", so they flew patrols at 40,000' with all guns loaded, paralleling the imaginary line in the sky.

It was a thrilling thought to be up there and hear; *"BOGIES, THREE O'CLOCK LEVEL!"* and to see two MIG-15's flying in the same direction some five miles over. Or: *"BOGIE, 9 O'CLOCK LOW!"* and see a lumbering 4 engine "BEAR" on a photo mission.

Nothing ever happened, but the thought of maybe getting involved in the first fire exchange of W.W.III, made the adrenaline flow a lot faster.

After a few days, Matje got the O.K. for the Dutchmen to fly together and from there on in they did just about what they felt like, weather permitting. The weather briefings at 8 am were tough on hangovers, but otherwise a lot of fun.

The briefings were conducted in English as a courtesy toward the foreign visitors, but the METEO officer did not have the advantage of $1^1/_2$ years in the U.S. and had a terrible accent. He must have thought that the pilots really liked him, because they smiled at him a lot. The truth was, when he pointed at the map and said; "FOK" over Germany and "FOK" over Holland, it was impossible NOT to laugh.

One of their favorite sports was "HIGH ANGLE BOMBING" over a huge rock at the very Northern tip of the Danish mainland.

From 10,000' at near stalling speed below 200 MPH, they'd roll their jet over and onto its nose, going straight down for the rock. Speed would build up fast and of course as the plane gained speed, the nose wanted to come up. That required forward pressure to keep that from happening and keep the sight locked on the target. Theoretically, when the forward pressure on the stick built up to a negative $^1/_2$ G, the bomb would release automatically and the sight would go off, telling the pilot: "Bombs gone, pull out!"

Practically speaking, that had failed sometimes in Korea!

Pilots waiting for their sight picture to disappear would find out too late that they were too low, going too fast to pull out and splattered themselves all over the landscape.

The thing to do was, don't wait for the bombsight, when you reach 400 MPH, push the bomb button and pull like hell!

If they started pulling out at 400 MPH standing straight on their nose, they'd reach 550 MPH at the bottom of their curve.. That was a safe margin! Just a few more seconds and they'd pull out at ZERO feet, doing 600 MPH. That was too close!

A 550 MPH pull-out would have them right back up at 10,000′ doing 200 and ready to do it all over again. A fun job but a back breaking one. Pulling 5 "G'S" consistently coming out of the dive was rough on the body and murder on the arm muscles.

Great fun just the same!

At one point Matje pulled something that Frits had not expected from his conservative flight leader. They were right on the deck over the inland sea called Skagerak coming up on the mainland coast. Being right on the deck over water in low level battle formation was not all that unusual, but Matje didn't pull up when they reached land and with Red on his wing he flew right into the harbor, staying just 5 feet above the surface. Of course Henk, Frits on his wing did the same, and at 420 MPH they flew right in between the ships docked and moving in the harbor!

The crews on the decks looked down on them! It was fascinating and exciting. Frits had to stay in position on his element leader and at the same time keep from clobbering into something and ready to pull up at a split-second's notice.

They had a ball!!!!

Social life was something else. As in America, all pilots were officers, so they each had their own little suite in the B.O.Q's.

Luxury plus! Fit for visiting Generals or Dutchmen.

The food in the Officers Mess was outstanding, especially the noon hour SMORGASBORD. The best of Herring, Caviar, Salmon and dozens of other varieties including shrimp were there for the picking. A great daily treat.

The Danish beers, "CARLSBERG" and "TUBORG" were most tasty and their liquors "AQUAVIT" and "BRUN HERRING" were outstanding.

The nightlife in town was something else! The town of KARUP was about 30 minutes from the base and the Danish pilots were great hosts. They always provided transportation and they also taught the Dutchmen something about Scandinavian "FREE LOVE". Years later, Frits still had a hard time believing it himself when he related the stories. It was as if it really hadn't happened to him, but read about it somewhere!

First the Danes accused the Dutch of being narrow minded about sex, just like the Americans.

"We Scandinavians look at sex as just another body function to be enjoyed, like eating, drinking, exercising or dancing! Not as something dirty that needs to be hidden like you Puritans do!"

"We? Puritans?"

"You'll see. You'll meet girls, ask them to dance, buy them a drink and they say yes! Yes? O.K.? You'll see! When she takes drink, she'll have sex with you. When she buys next round or refuses drink, she won't have sex with you! You'll see!"

Well, they did see! As a matter of fact they did a lot more. They had been in a club with an impressive name like; "HANDWERKERVEREIN" or something like that and had landed in a club called: "BLUE MOON", an even more unlikely name.

From the bar, Frits had spotted two girls at a table and asked one of them to dance. She spoke no English at all, but a little bit of German. At least they could get some words across. They didn't need many words, cause when the little combo struck up; *"Into Each Life Some Rain Must Fall!"*, they did a modest "LINDY" together and hit it off rather well.

He helped the music along with Ella Fitzgerald and Ink Spots imitations; *"But Too Much, Too Much Is Falling In Mine! Into Each Heart Some Tears Must Fall!"*

She was enjoying herself, swinging her cute little body around the dance floor while smiling at him from ear to ear.

Nice going! When he brought her back to her table and said *"TAK"*, she must have thought that he understood some Danish and ran off a bunch of her "goobly-gook", of which he didn't understand a word. He girlfriend at the table didn't do any better but they did understand *"HOLLAND"* and *"HOLLANDER"*.

Her eyes held an invitation for which no one would have needed an interpreter, so he retrieved his glass from the bar and joined them.

They accepted a drink from him when he offered!

He couldn't help but think about the words of his Danish "sex counselor". *"TWO , NO LESS?"* Impossible!

One of the Danish pilots joined them and danced some with the other girl and meanwhile talked that funny language of theirs. Danish had Frits stumped. In print he could figure it out, but spoken? It all sounded like; *"Oy-roy-roy-oy-oy-roy-roy!"*

Couldn't understand a word.

The Dane turned to Frits; *"Nice girls, yes? You like her?"*

He was looking at the one Frits had been dancing with.

"Yeah, she's kinda nice!"

"You wanna fuck her?"

"Oh, please, not so crude!"

"That's o.k., I'll ask her!"

"No, no, come on! No, no!" Frits looked embarrassed.

"Puritan!" The Dane turned to her and spoke some of his; "Oy-roy-roy-roy!" Again.

She smiled at Frits and the Dane said; *"She would like to!"*

Frits figured he was being had. The Dane probably asked her if she liked the way he danced or something, but he figured he'd humor the clown and said; *"Thanks, great!"*

Her name was "BRITTE", short for "BRIGITTE" and she and Frits just danced the night away. She seemed as entranced as he was. From the Dane he learned that the two girls were nursing students and lived in a dorm on the top floor of the Hospital where they were in training. So much for a chance at some privacy! Well, he might as well enjoy the dancing and the necking during the slow ones.

At quarter to one the lights came on and it was time to go. The little SAAB was packed. Frits had Britte on his lap in the backseat, her girlfriend sat on another Danes' lap in the frontseat and one of the Dutchmen sat next to Britte and Frits in back. Six people in a little 4-seater. Frits wasn't even aware of where they were going, he was too busy and so was she. Their hand roamed rather freely in that narrow backseat.

When the car stopped and the girls got out, Frits started to follow them but was told; *"Sit back and wait!"*

"Wait for what? I wanna say goodnight to her and I.."

"She'll be back! They just went to get a few more girls!"

"They're getting a few more girls? From where?"

"From the dorm of course, you dummy! There are a hundred single girls up there and we need only two!"

"You really believe they'll be back? Come on! We've been had! Let's go to sleep!"

"No,no,no. I told you, you dont know Danish girls!"

"You mean to tell me," he was in awe, *"that these girls right now are going down a row of beds, shaking the nurses awake and asking them: Do you wanna fuck?" "No!" "Go to sleep!" Or "yes!", "Get dressed, let's go!" "You believe that?"*

They laughed out loud and said; *"You've got it right, that's what they're doing!"*

"Sure and I'm the Queen of Sheeba!"

Denmark after an exciting flight

"Well, move over Queen, cause there they come!"

Sure as hell, four girls rushed out and crawled in all over the guys!

Their was an awful lot of giggling, patting and *"ROY-ROY-ROY"* going on as the car drove off, direction: base!

Frits had his Britte back in his lap and they took up where they had left off; kissing, exploring and getting excited!

Britte damn near raped him in the car and they had to all straighten up as they approached the gate, where the chauffeur just hollered; "HOLLANDER" and the gate opened.

Because each Dutchman had his own private suite, it worked out like a charm. Each suite had a bedroom and a living room with a couch and Frits generously gave up his bedroom. The couch was big enough for what he had in mind.

Then something went dead wrong! Instead of finishing the rape scene that she had started in the car, Britte was upset, talked *"ROY-ROY-ROY"* a mile a minute, wouldn't co-operate and finally knocked on the bedroom door. More *"roy-roy-roy"* and Frits was told that she didn't want to be on the couch with Frits, she wanted to be in the bed with them!

"OH?, OH??? WELL, NO SWEAT! or rather *"LOTS OF SWEAT!"*

He tried to screw both of their brains out! Nearly did!

At 4 in the morning, the proud SAAB owner woke them; *"We got to get these girls out of here!"* Frits helped. He rode back and forth with him, delivering the girls back to the Hospital. The other lazy bastards stayed in bed!

What a night! No wonder these guys killed themselves flying; *"THEY COULDN'T KEEP THEIR EYES OPEN!"*

The Danes proved to be excellent hosts. They treated their guests to a TOURBOAT cruise to the highest point of Denmark, 110 feet high and a visit to "COBNHABN" for the weekend. A transport flew them in on Friday night and back to base on Sunday night. One of the young Danish pilots had a MAM with a huge house and she welcomed all of them to stay with her.

What a city!!!!! COPENHAGEN, "COBNHABN", as the Danes called it was a beautiful city with plenty of sights to see and lots of places to go. The "TIVOLI GARDENS" alone could keep someone entertained from 9 AM till 2 AM without letup. There was indoor and outdoor entertainment all through the day and night. Acrobats, singing groups, dance groups, outdoor concerts, numerous restaurants and several nightclubs. They spent the entire Saturday there. It was just grand! On Sunday, they toured the city and with their pilot guide, born and raised in COBNHABN, they had a phenomenal tour. It was with regret that they had to leave.

From every viewpoint; flying, partying, sightseeing and eating, two weeks in Denmark was an unforgettable adventure!

CHAPTER XX
FUN AND FLYING!

Their flight across Southern Jutland, Germany and Holland at 1000 feet was like a rest cure. Straight and level all the way. If only they had had an automatic pilot, they could have gotten in an hours sleep. Compared to the way they had raced their aircraft across the skies for the last few weeks, this was like a snooze at a Sunday picnic.

It was warm too. Their luggage was stored in the gun bays and the battery compartments but, for a lack of space, they wore their heavy flight jackets with the phony fur collars. No need for oxygen masks at 1000' so they were left dangling from their helmets.

On Saturday afternoon, Eindhoven was deserted, so Matje called for a low pass or two and got clearance. On the first pass they came diagonally across the runway and down on the deck in box formation over the tarmac in front of the tower.

The weather co-operated beautifully. A high overcast, hardly any wind and no bumpiness. Matje could look the tower operators in the eye as they came thundering by and Frits, sitting below Matjes tail in the slot position, had to look up to them. Roaring over the 314 buildings, they pulled up in a slow barrel roll and made a 180 degree turn at the top. This time they screamed on by in echelon formation and heard Matje ask for clearance for a "pitch-up" landing. Looking good, hanging in tight, they flew the length of the runway on the deck.

"ONE SECOND!" was all Matje said and "POW", he was up and over, Red one second behind him ending up upside down at 1200' going in the opposite direction. "POW", Henk off and one more second, Frits pulled hard with his head all the way back looking above him at Henks tail. *"OH SHIT!"* His backward head movement caught his helmet on the fur collar and shoved the helmet forward over his eyes! Just for a second, he couldn't see out, relaxed the stick pressure kinda naturally and slapped his helmet back with his left hand. *"DAMN!"* The damage was done! On downwind leg the spacing between #'s 1, 2 and 3 was perfect, but #4 was twice as far out, ruining the picture. *"DAMN, DAMN, DAMN!"* He caught up some by flying downwind faster than the others and cutting the base and final turn tighter, so at touchdown the spacing was O.K. again. Thank God, there weren't too many people on base, but those who had watched it teased the hell out of him.

There was some very good news for them on arrival.

Since it was the end of the month, the paymaster had arranged for their pay to be held for them at Base Ops. Besides that, Red and Frits received their final settlement from the States! Reimbursements for T.D.Y., dry cleaning, etc., totaling about $75 or nearly 300 guilders. That was a nice surprise. The first thing Frits did with his money was very sensible; he bought 4 new tires for his car. Next, he settled his tab in the Mess and the "Poort van Cleef" with "Fat Wim" so he could start a new month with a clean slate and a clear conscience. He couldn't buy a money order on Saturday to send to Mama, but he put the money in his sneaker in his room. That way he wouldn't spend it, he separated his assets from his liabilities.

Next stop; the "CINZANO BAR". A while back he'd met this "MARILYN MONROE" type girl in a cute little corner bar, "THE SPINNINGWHEEL". It had been early in the evening when Ike and Frits explored the scene in the bar and were still on their first beer when a blond gal, about 23-24 emerged from the ladies room and joined another gal at the far end of the bar.

"I'm in love, Ike!"

"You're always in love, you shithead and then the next day you chase after some other tail, so dont give me your shittin' love stories!"

"I gotta check it out!"

Her blond hair was done up kinda "MONROE" style and she wore a tailor made two-piece black suit over a white silk blouse. Very decent looking, but very stunning and contrasting. When she walked to her seat, she wiggled somewhat as if she were tipsy and she was. Beer in hand, Frits moseyed over there and he had barely said; *"Hi, ladies!"*, when the blonde with a crooked smile asked; *"Hi, a pilot I suppose?"* Frits looked himself up and down and came back; *"Where do you see that written?"*

"It's all over you honey! Stamped on your forehead!"

She was nicely tipsy alright! Her tongue doubled a little and she giggled too much. But, she was in fine shape and that's what counted. There was an old Dutch saying; *"A drunk woman is an angel in bed!"*

"I *know you clowns, that's why!"* She rattled off a few names and some rang a bell. *"Right, I know Hompie and Fons, they're at Volkel, but the others don't ring a bell. Do we all look alike?"*

"No, but you're all kinda cute and cocky and all of you think that you're God's gift to womanhood and you're not! That's why!"

"Well, so much for good first impressions. My name is Frits. And yours?"

"This is Jo, pointing to the girl next to her, *"and I am Beppie and our cab is here, come see me sometime in the Cinzano bar. Bye!"*

"Have a nice evening!", He shouted after them and they were gone!

"That was the shortest damn lovelife you've ever had!" Ike was in rare form tonight. *"Hup, drink up, hup, it's paid for. Let's get outta here!"* Ike could consume beer faster and hold it longer than any person he had ever known. He had grown up in the beer distributing business and could really handle it. His Dad was the "GROLSCH" beer distributor in Deventer and Ike had learned how to handle beer trucks since he was fifteen. He had the damnedest habit of hitting a bar, drink one beer at his leisure, rush down the second one and say; *"Hup, let's go!"* All a guy could do then was gulp down his beer and follow.....

This time Frits was driving so Ike couldn't rush him too much.

Getting a drivers license in Holland had been a chore and a half. Frits had been well forewarned. His brother Joop had failed the practical driving test six times before passing on the seventh. Shortly after settling down in Eindhoven, Frits went

F-84 leader

to the nearest driving school and told the owner-instructor; *"I know how to drive, but I'm willing to forget it for my test. I want to learn in your car, not mine, how to pass the test. Fair enough? You just teach me how!"*

As silly as some of the rules were and the pointers the instructor gave, they still spent two hours rehearsing. The most tricky part was; parking along a curb on a slight incline, wheels turned toward the curb to keep from rolling, the handbrake set, ignition turned off with the shift in gear. Along the same lines was another "must". Stopped on an incline, the instructor would get out, put a matchbox behind the rear wheel and the student had to pull up without slipping back one quarter of an inch, because that would crush the matchbox. Crushing the matchbox meant automatic failure. After two hours of that, Frits took the test, made no mistakes and had his license in twenty minutes. For years he ribbed his older brother about that feat.

He had never been to the Cinzano Bar before meeting Beppie and there was good reason for it. It was a neighborhood bar, far removed from all the pubs in the center of town. When he did go, he found a surprisingly neat and well lit bar, kinda like "RUSTWAT" at home, but with just one billiard table.

Her reaction surprised him; *"It took you long enough, what kept you?"*

"Well, I'm glad you remembered, Beppie. How are you?"

She knew how to make a guy feel good. She was working behind the bar and not tipsy this time. Her Mam, a much older version of Monroe, was sitting on the corner stool handling the cash register and "JAN", a pipsqueak of a man with a huge scar across his face was waiting on tables. He was Mam's husband and Beppie didn't show a hell of a lot of respect for him by introducing him as "JAN" rather than "PAP" or "DAD".

Mam was very friendly, but in the bar business, that might all be a veneer, as he well knew.

He was flattered that Beppie even remembered his name and he learned that when they met, it had been her afternoon off and she and her friend had caught a movie and had a few drinks.

"Too bad I had to get back to work that day. I was feeling no pain and was in a great mood. Too bad!"

When they heard that Frits' Mam also ran a bar up north, they had to know all about it and within the hour they had heard his entire family history. And he had heard theirs.

Beppie had gotten married at 17 and had a 5 year old boy named: "LEXIE", short for Alexander. Her life apparently revolved around him and the whole family lived happily together over the bar. Frits shot some billiards with Jan, bought a round or two and was on his way after a few pleasant hours. Her next evening off was not until the next Monday, but they made a tentative date any way, with the promise, he'd drop by before then.

On Monday, he picked her up and went to eat in a nice little place in a little village called BEST, just outside of town. They ate and drank and talked at their leisure and afterwards stopped by the Chaplain's place, the Tjitjac, on the way back.

Being that it was Monday night, there was no other place to dance anyway. The place was quite crowded and noisy and they shared a lot of handshakes and drinks. They were getting along beautifully and had had a great time. As a matter of fact, they got along so well, that on the way back, Beppie was nearly sitting on the driver seat with him, kissing him feverishly and Frits had his hand inside her bra, hanging on to a well proportioned boob. He wasn't going very fast, maybe 30 MPH, but when a front tire blew, he reacted wrong, slammed down the brake and skidded into the ditch before he knew it.

The Ford Anglia after it hit the tree

Beppie bounced against the rearview mirror with her forehead and blood was all over her face and blouse. It happened right across the street from an Army camp, the guard saw it happening and within seconds, they were being helped out of the car. An army ambulance took Beppie to town while Frits waited for a tow truck to pull him out. The damage was not extensive but the car was not driveable, so Frits found himself walking back to the barracks.

The hospital emergency room informed him on the phone that Beppie had been treated and released and he felt relieved about that. He didn't have the nerve to call her home. He figured Mam would wanna kill him. Things like: *"IRRESPONSIBLE BEHAVIOUR"* and *"RECKLESS DRIVING!"* etc..

The following evening he reluctantly made the trip to the Cinzano Bar, knowing damn well that "MAM" would have his hide and lace into him. He was fully prepared to accept the fact that Mam nor Beppie ever wanted to see him again. As a matter of fact, he wished that he were more of a coward, so he could simply stay away and not even face the two of them at all.

He had bummed a ride and walked into the nearly deserted Cinzano Bar at 6.30 and found Jan and Mam smiling at him; *"You had quite an interesting evening, didn't you?"*

Talk about relieved!!!! *"How's Beppie?"* was all he could say.

"Oh, she's fine! She called about a minute ago that she'd see you at the Poort van Cleef at 7, so you'd better hurry!"

He was flabbergasted! He wanted to ask; *"How about her head? How about all that blood?"*

but he didn't.

At the POORT she was at the bar in the same ladylike black suit that she had worn when he first saw her.

"Hi honey! How you?" She was already tipsy. She kissed him smack on the lips and laughed out loud;

You have a hell of a way to treat a lady! I had to go home by myself!" She laughed again at her own joke. *"Have a beer! I took another night off and I'm treating!"*

"But how's your head?" He was still worrying.

"Oh, I took an aspirin this morning but I'm not sure whether it was because of the cut in my head or because I had a hangover. Anyway I'm fine and you gotta quit looking so worried! You look like you need to lay an egg! Do you know my friend "Wim" here?" She giggled again. *"Nice work Frits!"* Wim was no help. *"You take a lady out for the first time and she lands in the hospital. I hope you fly better than you drive! Remind me never to drive with you!"*

"Screw you and give me another beer, doesn't anyone want to know about my poor car? How's it doing? How much damage? How long will I be without it? No sympathy from any of you bastards! I think I'll go sit in the corner and cry in my beer!"

Beppie jerked him closer to her, *"You'll do no such thing! My sister Truus is waiting to feed us and you're holding things up. What do I owe you, Wim? And call us a cab, please!"*

So much for major injuries! Just a cut with a lot of blood apparently. A band-aid covered all the damage. Thank God for little miracles!!

Over "biefstuk" and asparagus, Truus wanted to know; "*How'd you get that cut anyway?"*

"She hit the mirror. It went all so quickly! We were driving along and "pow", we're nosedown in a ditch. I didn't know we'd had a blow-out until the towtruck pulled it out of there!"

"She hit the mirror? Where were you sitting?"

"Hahahahaha, I was trying to sit in his lap and I nearly made it! Hahahaha" Beppie thought it was hysterical. Truus was shaking her wise old head. Truus was Beppie's senior by 6 years. Unlike Beppie she had black hair and brown eyes.

She didn't resemble Beppie nor her mother. Her husband of ten years had recently been killed in a motorcycle accident on his way to work one morning and left Truus with a 9 year old boy "Wimpke" and an 8 year old girl "Truuske" to care for.

She ran a little grocery store and eked out a meager living.

Her house was small to start with, but by having the front room converted into a grocery store didn't leave her much living space. She was hacking it though and that's what counted! The death of her husband had crushed her and she was grateful for some company and diversion.

By 11, Beppie called her mother that she was spending the night with Truus and by 11:30, they had the couch to themselves.

They acted like they had been denied sex for the last ten years and when the taxi delivered Frits to the base the following morning, he was just in time for the 8 AM briefing, but couldn't see straight enough to identify one single plane.

Good God, he needed to sleep somewhere! That's how they got started and that's how they continued! One common interest! SEX! Little else! But it seemed enough.

So, after two weeks in Denmark, he was on his way to see her, new tires and all.

"Hi, honey! What did you bring me?"

That was all she could think of? *"Wouldn't you rather ask the "love of your life"; how were your two weeks? Did you miss me? Are you o.k., honey? No, just; "What did you bring me?"*

"Oh, you poor dear!"

She snuggled her voluptuous figure up against his and his temperature rose by a degree per second! Boy, the things she did to him!

"That's all?, A lousy set of earrings? Cheap little junky earrings with the name of a movie house on them?"

The proud founding fathers of Copenhagen were turning over in their graves, Frits knew, when they heard their famous "TIVOLY" compared to an ordinary movie house.

"Well, maybe I better leave again and see if I'm appreciated somewhere else!"

"Like hell you will! Listen to the new records I bought and Jan, why doesn't Frits have a beer yet?"

"Boom-Sha-Boom, Tjalala, Boom-Sha-Boom, Life Could Be A Dream, Sweetheart! Hello, Hello Again, Sha-Boom, I Know That We'll Meet Again.

Beppie and Frits

"Let's roll up the carpet and swing!"

"TJALALA!" Saturday night in Eindhoven! *"Life Could Be A Dream, Sweetheart!"*

Life could go on like that forever and it nearly did.

Mam sat him down and made him blush one quiet Saturday morning.

"I know you guys do it and do it all the time." She raised her hand, *"Don't say anything, no point in silly denials or explanations. Whether you do it in the car, the wheatfields, at Truus', or in an alley, I know you do it and I can't stop you guys. So instead of doing it where you might get caught or embarrassed, I want you to stay here! So at night, you leave the bar as if you're going to the base, park your car around the corner, out of sight because we don't want Jan to know. Then come around back through the kitchen and crawl in Beppie's bed so you can get a good nights sleep for once. Man, you've been killing yourself! Besides that, your hanging around in the bar all the time is bad for business, so at night, just say goodbye, come around back and crawl in bed. When she's finished, Beppie will wake you, I'm sure! I'm not worried about that! Now go somewhere! Goodbye!"*

Holy Moses, what a speech!

Well, from there on in he'd hang around with the guys until a reasonable hour, drive to Beppie's and climb up to her room via the flat roof over the kitchen. Nuttin' to it! Jan had locked the kitchen door on him a few times and couldn't understand who might have unlocked it.

Bep would join him at 1:30 or 2 o'clock and at 6:30 he would head back to the base and be at the briefing all bright eyed and rested, even recognizing the silly aircraft silhouettes.

A few changes came about at the base around that time. All "Non-Com" pilots were moved to Kamp Zeelst on the opposite side of the base and a new Officers Club and B.O.Q.'s was opened right across from the main gate. An actual "SPITFIRE" decorated the front lawn of the Mess and indicated the mentality of the senior officers; "THEIR HEARTS WERE STILL IN THE PROP-AGE!"

Tragedy struck again!

Soesman had led a flight of 4 on a low level mission over Germany and about 20 minutes from Eindhoven, near Apeldoorn, his #4 man, Pierre Gilliam ran out of fuel. He flamed out and belly-landed in a freshly plowed potato field. The plane was nearly unscratched, proof of a fine flying job, but Gilliam broke his back and was hospitalized.

Frits was fuming; "*THAT STUPID ASS SOE-OESMAN!*"

The flight patterns itself after the man with the least amount of fuel, which is always #4, and gauges its return to base on #4's fuel status. They had passed within just miles from "TWENTHE AFB" and were within minutes from "DEELEN" when Pierre went down.

"SOESMAN SHOULD BE QUARTERED BY FOUR HORSES!"

That was not just Frits' opinion, a lot of guys shared that. Nothing came of it, though and Soesman remained OPS-Officer.

Those who had criticized him just got to fly less and his cronies got to fly more!

The Air Force started something new! They started their own Officers training! Every 5 weeks a 10 week course would start, called; "O.R.O.V." (Opleiding Reserve Officier Vlieger). Of the 314 bunch, Red was the first to go, he joined the class of O.R.O.V. II.

Frits was sent to Nijmegen for a Psychiatric Officers Evaluation and he decided to make the most of it. He hadn't danced in that town for years. Nijmegen was still a first class town and it was good to be back and enjoy! The faces had changed, but not for the worse, they could still swing!

The next morning he went through the same black magic and mumbo jumbo that he'd been put through 3, 4 years ago, at which time they determined; he "HAD" it as a pilot.

This time the evaluation was about the same and whatever these clowns called "Doctors" saw in him and his answers, he agreed with them again; *"You'll make an excellent pilot-officer, a good leader in the air. But..., From a military standpoint?...A fairly lousy officer!...All you need is a war and you'll be a young general if you survive. In peace time???? I don't know?"*

"Well, thank you sir, I'll take that as a compliment! He was truly flattered and felt; *"I could have told them that!"*

"SHORTY" had been transferred to BASE OPS. The story behind it was that he was too big for a fighter and if he had been an American Cadet, he would have gone to bombers. It was a fact, indeed, if he would have had to eject from an F-84 and made it, he would have lost both legs, just above the knees. Now he flew T-6's around and was being checked out on the old C-45 that was still around. It represented Holland's entire bomber force. He was in no way depressed over it and quote; "FLEW HIS ASS OFF!", playing messenger and delivery boy.

He'd been up at Leeuwarden delivering the payroll when the squadron was up there for Air to Air.

He was still the best "musical imitator" and his "Bass Fiddle" imitation with his big tongue flipping in and out, still left people in hysterics!

In the Cinzano Bar, he would at popular request mimic a crippled helicopter coming in for disastrous landing. He needed the entire pool table for his crash while his arms were wildly windmilling over his head. By the time he rolled off the pool table, he'd have the customers howling with laughter! They kept buying him beer and he kept on performing.

In the Dance Club "BENELUX" next to the Poort Van Cleef on the Market Square, he was holding court most evenings. In full uniform, proudly displaying his two wings, wearing a little multicolored ski-cap and standing on the foot rail of the bar, he'd tower over everybody by at least a foot, taller than anybody in the place. There he'd stand and direct; " *Let Us, One More Time, Sing The Beautiful Song For All Who've Fallen.... One...*

Two... Three; "Hela"...." and there they'd go again...! There must have been 15 of them, mostly in civvies, whooping it up with another 100 customers in the joint making for a rapt audience.

The moment the band would strike up the first note for the next dance set, he'd holler: "*CHARGE!*" and the whole gang, followed by a dozen girls would run out on the dance floor and sit down in a circle surrounding "SHORTY" who would lead them into another song or the "SQUADRON TOAST".

The toast had been brought back from Texas or Arizona but nobody quibbled about its origin, they just shouted it out!

"HERE'S TO ME WHEN I'M IN A SOBER MOOD
AND I RAMBLE, SIT AND THINK.
AND HERE'S TO ME WHEN I'M IN A DRUNKEN MOOD
AND I GAMBLE, SCREW OR DRINK!"
AND WHEN ALL THESE MOODS ARE OVER
AND FROM THIS WORLD I PASS, I HOPE, THEY'LL BURY ME UPSIDE DOWN
SO THE WORLD CAN KISS MY ASS!"

Then with a *"ROAR"* and a *"PROOST"* they'd jump up and make it back to the bar to continue their shenanigans. Nobody ever threw them out! Maybe management enjoyed their nonsense too or maybe they were good for business. They sure as hell dropped a lot of money in the place and they attracted girls wherever they went.

They thought that life was meant to be a long and continuous party and it certainly shaped up that way!

On November 14, Frits invited the gang over to St. Pancras for his birthday party at his Mam's place and he called "Volkel" to leave word for his old pals like Henk Koppelman, Frits Flieger and Jan de Ruiter, assuring them; *"Don't worry, we'll find you a place to sleep!"*

What he didn't know was, how word got around. He had told Mama to have enough beer on hand and that he was paying.

Mama's answer; *"To you, it's wholesale!"*

It was a good thing she said that, because at 4 on Saturday afternoon the first guys started drifting in already and although it started slowly with a game of billiards and some hot flying tales, the spirits rose quickly.

Some of the local beauties must have gotten wind of the affair and they floated in and soon floated around the floor in tune to the jukebox.

In order to keep easy track of the out flowing beer, Mam served Frits' guests bottled beer, so all they had to do was count the cases.

By 7 o'clock, seven pilots were present, some local girls and a few regular Saturday night customers, plus of course, the family! Quite a party already!

By 9, there were at least twenty pilots and by 11, Mam counted 35. One car with five of them had come all the way from Leeuwarden across the "DIKE" and one carload of "322" Pilots and girls from Soesterberg had heard in "REYNDERS" in Amsterdam that there was a party in North Holland and that "ALL" pilots were invited.

Mama was doing a whale of business, because only the beer was on Frits, so all the other booze, soft drinks and food was out of pocket. Joop, Clasien, Paul and Hedwig worked their hiney's off, cooking and serving. They did get to laugh and participate some though!

In the small town of St. Pancras, Township Koedijk, closing time was 11 PM. So when the entire police force, both of them, showed up at 11:30 there might have been 100 people in the joint. The cops considered Frits the "town hero" and so they hung around for about half an hour, refusing dozens of drinks, trying to get an exodus started.

"HE'S A JOLLY GOOD FELLOW!" and *"LET'S HAVE ANOTHER TOAST FOR THE POLICE FORCE!"* didn't help matters at all.

The following day Frits heard about the secret deal Mama had worked out with them.

The deal was: the doors would be locked, no new people allowed in. That would make it a private party and the Police would reappear at 1:30 to help Mama shove everybody out of the door. Mama was an old hand at that type a thing. Plus the cops had a lot of respect for that old lady, now 49. They had worked together many a time.

The hootin' and hollering, the dancing and singing, the hugging and kissing continued full blast until 1:30 without any signs of letting up nor any signs of anyone leaving! So, at 1:30 the doors swung wide open, the jukebox was cut off and Mama started collecting the outstanding tabs.

Little by little, with more singing and more kisses, people started to leave and by 2 o'clock only the overnighters were left over and most of them helped clean up the place while drinking more beer.

Frits Flieger had already been put to bed in the attic and mattresses and box springs were all over the attic floor, making for makeshift beds. Just as everybody got ready to retreat, Flieger came down in his under shorts needing a leak and beer, *"IN THAT ORDER!"*

Mama decided to leave a lot of lights on, so the guys could find their way down, since their were no bathrooms upstairs.

The family was still up, sitting around the big round table settling Frits' tab. He didn't have enough money to cover the whole bill and had to promise to settle the account at the end of the month. His tab was three times as high as he had anticipated and yet that was at WHOLESALE COST.

All his brothers and sisters agreed that he had a bunch of swell friends and that most pilots were absolutely looney!

How looney, they found within the half hour, when someone screamed; "FIRE!" Frits Flieger had fallen asleep with a burning cigarette and had burned a sizable hole in the sheets and a blanket and the whole bed might have gone up in flames, had it not been for the fact that the flames burned his back. Finally, everyone got some rest.

The old homestead

Over breakfast and beer, they rehashed the adventures of the night before and came to the conclusion that some guys that "definitely" were going to be there hadn't showed, which caused Mama to sigh; "*Thank God, we had enough as it was!*"

Later in the week they learned that 2 carloads of guys and gals left a bar in "LOOSDRECHT" heading for St. Pancras, but lost their way and never made it!

"What do you mean, Mam?"

"I mean, thank God that we didn't have another dozen people show, we wouldn't have had enough beer left for our regular customers of today!"

That was Mama, forever the business woman! One thing Mama did agree with; her first meeting with all of Frits' buddy pilots had been a nice one. She thought they were nutty but delightful. All in all, a great bunch and a great 23rd birthday party. In her devout Catholic mind, some of the girls were a little "LOOSE" and "PLASTERED" and girls shouldn't smoke and drink like men, but overall, she had thoroughly enjoyed the crowd.

She had danced with some of the boys, she had laughed herself silly at some of their antics and nobody had tried to skip out on their bill. The only things that were stolen were some ashtrays, but that might advertise "RUSTWAT" in various parts of the country.

*"You boys be careful now when you fly those jets. Frits scares the daylights out of me when he comes over and goes like so...and so... and so,,,(*her hands were maneuvering like a fighter pilot's*) and so....and like so...., I don't like that!"*

"Fritsie-boy, you come and fly over the house sometimes?"

"Who? Me?"

CHAPTER XXI
BREDA

"*Report to Breda for* OROV IV."

"Who else is going?"

"From this squadron? Only you, Forrer!"

"Call'em up!"

From 315 there was only one as well, "Ger Walhout". Ger was a quiet guy who had been a class ahead of him in the U.S. and whom he'd briefly known in LUKE and then again in Eindhoven.

Ger had a car, Frits had wrecked his car again and it was in the shop, maybe beyond repair. Ger and Frits would ride out together in Ger's car which was a hell of a lot better than going alone or riding the train.

When Ger pulled up in the morning, Frits had the few necessities in a little bag, as no civilian clothes would be allowed in Breda.

The prospect of finally becoming an Officer started to appeal to him. His past record, including his time in the brig and the Texas jail, would be destroyed and he would start with a clean slate. After ten weeks in Breda, they'd be on some sort of probation for six weeks and then they'd be promoted to "Warrant Officer" and after 3 more months to 2nd Lieuteneant. It seemed like a stupid system, but they all agreed; they had no choice. By the time his American classmates were already 1st Lieutenant, he was being considered for 2nd. Sounded sick.

By the same token, if they were to stay in the service and wanted to continue flying, they'd be better off as Officers. The pay was better, they'd become flight leaders and they'd get more hours in. Besides that, they'd have to put up with less chicken shit. So, that meant for ten weeks he'd have to toe the mark. He should not screw up, he would do his best and work at being # 1 in his class. He had no doubts about that! Some may have called him conceited, he called himself self-confident. He knew that if he buckled down he could do whatever someone else could do. And with some additional effort probably better than anyone else.

With that in mind, they were off to OROV IV.

"BREDA" is an ancient city with beautiful parks, avenues and buildings. The city played a major role during the "eighty year war" between Spain and Holland. The Spaniards occupied this old fortress city for many years and left their stamp on its buildings and cultures. The best of the "DUTCH MASTERS CIGARS" were manufactured right there in Breda and have been for hundreds of years. The people of Breda are typically southern Netherlands, meaning open, friendly and fun loving. It had always been one of the best "Mardi-Gras" cities in the land.

That's where Ger and Frits arrived at the old military camp that now housed the OROV training facilities, as well as the NCO training complex. They arrived with all sorts of good intentions and were full of enthusiasm but were shot down immediately.

"I'm Captain van Krimpen and this is Lieutenant Herberts and we are in charge of you and your training for the next ten weeks!"

Van Krimpen must have been 5'3" with a high-pitched voice like a female, but he did have a blue "Korean" bar on his chest and paratroopers wings. Herberts must have weighed twice as much as van Krimpen. He was a good 6'2" and probably 240 pounds. It was rumored that he had washed out of the flight program and resented every pilot who "made" it.

"Here, you remove all your stripes, decorations and wings and for ten weeks you'll be like any other recruit. You will not be allowed off base for the next two weeks and then just sparingly after that! Bed check at night is 10:30 and you'll be either in your bunk, under cover or in front of it in full uniform. Classes start at 8 am sharp, right after breakfast. Reveille is at 6 a.m. and standby inspection at 6:30. Your first function is tomorrow and there will be a bed check tonight. Have a nice day, gentlemen!"

"DISMISSED!" added Herberts.

"Are they for real? Four years ago I went through this chickenshit and then twice more in the United States and now I've been a pilot for a year and a half and this is how we're gonna be treated?" Frits was madder than a pistol.

He wasn't the only one sharing that sentiment. There were 28 "Pilot-Officers-to-be" from all the different Air Force Bases plus a Marine candidate. The Marine was only 19 and was going through a special Marine program and would probably become the first Marine Pilot in Holland's Marine history, especially since the Marines didn't have an aircraft as yet. Derek, the young Marine, was the only one who didn't share their beef, because he had no reason to.

All the others had gone through training in the U.S. or England for Gloster-Meteors and were highly teed off.

Again: a big room with thirty bunk beds, lousy shower and toilet facilities, it was indeed like Nijmegen in 1951 and now again the same crap in 1955!

The only fun was: meeting all the other hotshots.

The rivalry between the F-84's and the Meteors had been an ongoing one and was continued, hot and heavy, in that 2nd floor room of the O.R.O.V. IV class.

From Volkel, a few familiar faces were present and a few new ones.

One who stood out was Geppy Steenhuis, a fat guy with a shock of blond curls. If you didn't notice his unusual size, you would certainly not miss his mouth!

He was from Amsterdam and acted as if that gave him superiority over all other human species, sometimes in a funny way, most times aggravatingly so.

Most pilots were trim and athletic and Frits couldn't help but wonder how Geppy made it through the military part of the training. Anyway, he was there and he acted like he was the hottest thing around that could handle a stick.

By 8 o'clock, there was little to do, but sit around, so Ger Walhout and Frits decided to go out for a beer.

Nobody stood guard except the one at the gate and he just

saluted as they drove out. Very few people owned cars in those days and only high ranking officers and pilots could afford them and the guards didn't question either kind.

Captain Van Krimpen, the Korean War Hero, rode to camp on his wife's bike and Lt. Herberts had at least a sporty mans bicycle.

The very first bar they walked into was a friendly looking pub on the square across from the old Cathedral. The very first bar was also the place where Lieutenant Herberts sat at the bar and said: "Hi fellas! Let me buy you a beer."

"Oh, shit!" Frits even walked in without a hat.

It was too late to do something about it, so they joined him, chatting amicably for a while and left.

"We better get in there before bed check, right?"

"Yes", Herberts agreed, "You better!"

He never brought up that first violation, but he might have noted it on their report card.

Life wasn't all that bad throughout the training. A lot of camaraderie developed and endless war stories were exchanged with the F-84 guys learning a lot of insight in the Gloster-Meteor and vice-versa.

Evenings together in their quarters were pleasantly spent between studies, debates and song.

Classes were generally interesting and it was just a matter of buckling down and learning the old military laws, ranks, history and regulations.

Kapitein van Krimpen, all 120 pounds of him, made an ass of himself during one of his classes by getting involved in fighter tactics, of which he knew very little.

Everyone took an attitude: "Just let him ramble on. Don't let him get under your skin!", until he made a real asinine statement

"The old adage: attack out of the sun, has become completely outdated with the advent of radar controlled sights!"

Well,,, there was no way in hell Frits was going to keep his mouth shut on this one and he didn't. Explaining how radar sights in a fighter only became valuable when locked in on a target, not for detecting a target, Van Krimpen still didn't want to budge from his position until nearly the whole class started backing Frits and only then did he retreat somewhat.

"I'll check it out with the authorities!" was all the dumb ass could come up with. Obviously, twenty some odd pilots, deemed "combat-ready" by the U.S.A.F. and the R.A.F. were not considered "authorities".

Work in the field was more fun.

Playing special forces games, being dropped off a rolling truck in the middle of the night on some dirt road without money, maps or compass and having to make it back to base somehow, made for a great challenge.

When Frits and Derek were the first ones back, their feat was denounced, because they had commandeered a seat on a train.

"What's fair in war?" was not a consideration.

The rifle range was old fashioned fun and that was Frits' forté!

Here he could show off!

Herberts challenged them individually to outshoot him and he shouldn't have done that, cause some of them did!

When he dared them to pull the pin on a hand grenade, count: "21, 22" then roll it down an embankment and look after it, "23, 24," then duck, "25" and "boom", he had to slap Frits' head down when he was still "looking" at "25."

Herberts didn't know that Frits had gone fishing with hand grenades before he was twelve years old.

When they had a 20 kilometer cross-country scheduled, Derek, the Marine with the crewcut and Frits again became partners.

The Marine was in great shape. He had moved directly from boot camp to officers training and Frits had never lost his touch for long distance running.

The course was mostly an obstacle course through woods and fields, across creeks and walls and the fifteen teams left at 3 minute intervals. Derek and Frits last.

"We'll be in first!"

They didn't hide their intentions for one minute and when Herberts pushed the button on his stopwatch, they were off like a shot.

Passing most of the teams wasn't too hard except for a few who weren't about to be beaten by the "OLD MAN" and the "BABY".

By the time they were out of the woods and at the edge of

town, they had passed all but three teams and the last two really gave them a run for the money. It was tight! But the "old man" and the "baby" beat them in the stretch to the gate!

"Bragging rights reserved!"

Inadvertently a competition started within the group. It seemed that Geppy always found reason to challenge Frits to everything he did or said. As far as Frits was concerned, Geppy was an egotistical fat slob with a keen mind, but that was all. No comparison otherwise. Yet, the competition heated up when the days wore on and a "class president" was required.

Then... . all of a sudden... .. bombshell... . or two!!!!!

"Walhout and Forrer! Get in and get changed! You're taking off right now! Tomorrow morning you have to be in Venlo for a funeral!"

"Funeral??? Whose?????"

"Lieutenant Matje Willems!"

"You gotta be kidding! Matje??? Impossible! Anybody! Never Matje!"

"Get changed! Dismissed!"

"Holly hell! Matje Willems, the precision pilot? The stickler for detail? The man who doesn't make mistakes? Impossible!"

All the way back to Eindhoven, Ger and Frits kept guessing as to what happened, but they couldn't have been more wrong, no matter what they guessed!

"DROWNED? He drowned? In the plane?"

It was beyond belief. Not till they hunted down some pilots in a local bar, did they hear the real story.

Weather was minimal. Green cards only. Pilots with 1,000 hours and 50 hours "actual" (in the clouds) instrument time only.

Visibility one mile, ceiling 300 feet. Individual take-offs and landings only. Few or no missions were flown.

One F-84 from 314 was scheduled for major overhaul at the FOKKER factory at SCHIPHOL airport near Amsterdam.

Matje was chosen. He'd fly the plane up and take the train back.

Piece of cake!

For security reasons, the "top-secret" radar sight controls

were removed from the aircraft, but that was standard.

Matje had gotten clearance for take-off, started his run perfectly according to the tower, but when he reached "take-off-speed", he stopcocked the throttle, ran off the end of the runway, through the fence, across the road and into the frozen canal, his head barely submerged.

He drowned in front of two frantic Airmen. They were cycling back to the base, when the F-84 shot across the road in front of them. They jumped into the freezing water and with an axe, they hacked the canopy to pieces in order to get Matje out.

They were too late! He had drowned!

A thousand questions:

"Why didn't he keep his oxygen mask on?

"Why didn't they "blow" the canopy? Why hack it to pieces?"

"How could water get into an airtight canopy?"

"Why didn't he pull the gear and ground loop?"

"Why was he going so fast?"

"Why didn't he just take off?"

On and on and on!

In the final analysis it turned out to be such an accumulation of errors that it was just astonishing!

To start with, when the radar equipment was taken out, the attachments to the static and pitot lines, (for calculations for airspeed and altitude), were taped up, but leaked.

Result: improper air-speed indication!

When Matje was at 120 MPH, halfway down the runway, his indicator read only 80 MPH. If Matje had pulled on the stick, he would have flown, but instead he believed he was going 80 MPH, too slow to reach take-off speed, so he aborted. Yet, he actually was going too fast to stop before the end of the runway. Next mistake: he should have pulled up his gear! Why not? Nobody knew!

Mistake #3: He didn't blow his canopy, he rolled it back! When he rolled down the embankment, his canopy shot forward and derailed.

Mistake #4: He fired the ejectors and all he managed was: he blew two holes in the plexiglass!

When he rolled back the canopy, he also rolled back the two little metal "hands", against which the rockets exploded to eject the canopy. Because the canopy had not fully slammed forward, the little "hands" were not in line with the rockets and they just blew through the glass without lifting the canopy. Now he had more openings to let in water.

Mistake #5, he took off his oxygen mask which would have kept him alive till rescue arrived, instead: he just drowned!

Unbelievable!

After they heard it 15 times, they still couldn't believe it.

That explained why the linemen couldn't "blow" the canopy with the external handle, it was already blown, so that's why they had to go at it with axes. All this time, the cockpit was just one measly inch below the surface. Of all the stupid ways to die, this had to be the worst!

On the positive side, they got to sleep warm unexpectedly and the next morning they drove off early to attend the funeral Mass in the southern part of Limburg, the province where they had so much fun the year before.

As people filed into the church, a lot of handshaking and back slapping had to be done and some new faces had to be introduced.

Noticeably was a tall threesome: "KATJANG WILDBERGH, JAN WILLEM MOOY and MAARTEN SCHUIT." Apparently, they had already established a reputation for themselves and in turn had heard a lot about Frits as well.

He wondered why?

In church he found out! When during a serene moment a little bell rang out on the altar, Katjang whispered loudly: "CHOW" and sent the whole military contingent into a laughing frenzy.

Later, as people hurried out of their pews and lined up to receive Communion, they got Katjang to join the line, believing they were handing out pictures of Matje.

He looked rather asinine when he got close and made an "about-face" with a sheepish grin on his lips. It was a little sacrilegious, but hysterical just the same!

After the funeral, they had a brief chance to get re-acquainted over coffee and off they went, back to Breda.

Just 5 days later: same story... . different city.

This time a newcomer, Kees Rog from Scheveningen.

His death was just as puzzling. He dropped out of formation

at 15,000 feet and just went down and down and down... until he exploded into a million pieces, creating a crater, the size of a 10,000pound bomb.

"Why??? How???? No answers this time!

The newspapers talked about last minute heroics, sacrificing his life trying desperately to miss a nearby hospital.

In reality, there was no indication that he was even conscious on the way down.

"Why???" "No oxygen???" A person could live and survive at 15,000 feet without oxygen! No radio messages! None! No indications of trouble whatsoever! The whole thing was one big question mark.

This time they made it to the N.C.O. mess at cocktail time and after some beers, they mourned Kees' death in World War II style.

Backs to the fireplace, they toasted the fallen pilot, chug-a-lugged and pitched the glasses over their shoulder into the fireplace. The whole ceremony turned into a drunken brawl with a lot of singing, but Kees would have wanted it that way!

The next morning they were at the extreme opposite side of the country, burying the son of a modest fisherman, who fished the North Sea, day after day, as generations before him had done for centuries.

Within a week word came: "Captain Schippers, 315 Commandant, killed!"

"Forrer, you're not going!"

"Walhout, he was your boss! GO!"

"What the hell happened now?"

"Captain Schippers was one hell of a pilot! What could have happened?"

A phone call cleared that up, somewhat.

He had flamed out on baseleg, just 30 seconds from landing and had to make a forced landing in a soft field right in front of the runway. He must have dropped in pretty hard, because he broke his neck, while the plane remained perfectly intact.

"HOTDAMN! Three dead in the first ten days of February! What's happening to our Air Force and our airplanes?

Are we going to be like the Danes?"

Questions, questions, questions! But nobody, including Frits had any answers.

The motto was: "LIFE GOES ON!"

Life certainly went on for some.

In a phone conversation, an excited Hedwig told Frits: "Mama has been dating! She's in love! I think she's acting like a lovesick turtledove!"

In a letter from Mama, she explained about a gentleman, named Jaap Groot fom "Ooster Blokker". (a little village along the old Zuiderzee)

"He's a good Catholic! (important), a widower (important) and about the same age. (important)

He has nine children, (NINE?) mostly grown. Just two girls at home, 13 and 15." (Cute?)

Mama explained how he came acourtin' and how apprehensive she was even though it was more than three years ago since Papa died. He kept coming back and he was really a God fearing, (important) nice (important) man. (important)

Mam would like Frits to come home on the weekend and meet him sometime.

A letter from Joop was not so pleasant!

He wanted a "conference" as if he wanted to put a stop to the whole affair… .. NUTS!

Well,,,, the conference took place. Frits came home for the occasion and also had his chance to meet "Pa Groot."

"Pa" was a short, curly haired man, obviously born and reared in Noord Holland. His expressions and dialect were a dead giveaway, but he mixed it with a great sense of humor and great personality. It was impossible NOT to like the man!

Joop thought he was too simple, uneducated, unlike Papa, etc, etc.

Hedwig, the youngest one present at age 17, set his ass straight in a hurry!

"Who are you to talk and object? Are you going to be there when your Mam is lonely and grows older? No, you'll be away and married with your own problems and so will we. What do you want your Mam to do? Stay by herself and mourn Papa for the rest of her life? Or are you gonna keep her company?"

She was on a roll… ..

"Your Mama should make her own decisions without interference from us and she's entitled to every happiness God can give her! Heaven knows, the woman has been through enough, six kids, depression, war, welfare, sick husband, money problems all of her life! If she can get a bit of love and tenderness in her later life, we should all be FOR it, not AGAINST it! So there!"

She sat down.

"Amen!" "Amen!" "Well said, Hedwig!" "Good for you, girl!" Everyone agreed. Joop mumbled explanations and excuses, but the conference was settled.

"Good for Mama, good for Pa Groot!, Let's have a beer!"

As a matter of fact, Pa played piano and accordion (and the organ in church) and played them very well. Result: The group sang and laughed their hearts out, while having a beer or two!

Back in Breda, things were drawing to a close.

For weeks, the guys had been partying in a bar directly across from the "Gilze-Ryen" Air Force Base. The bar had been a pilots hangout since the 1920's and it was still their main meeting place. It was just 10 miles from Breda and right near the first train stop on the way to Eindhoven.

As classes were winding down and a final party was planned, that bar, the "Halfweg" (halfway) was chosen for the occasion.

It was decided that the gang would go by train, no driving, and that the class would invite the kapitein and the luitenant as their guests.

It was going to be a dinner party to end all parties, without girls though and only lasting till 11:20 in order to catch the last train back. It sounded great and just one last minute decision had to be made: The Class President!

If anyone but Big-Mouth Geppy had run, Frits would have bowed out of the competition and let somebody else have the honor, but "Geppy" to him was unacceptable.

SO,,,, Frits beat him by a majority of 25 to 5 and so he got to march the team to the railroad station, leading the singing: "*Sound off,* ONE, TWO, *Sound off,* THREE, FOUR, *one, two, three, four… . one two-a, three, four!*

My shoes are loose, my pants are tight, my balls are swinging from left to right! Sound off!"

On the streets, the platform, in the train, back on the street, they sang and laughed and laughed and sang!

In the private dining room, joined only by Lt. Herberts for dinner, the toasts and jokes went on and on and after dinner in the bar they tried to establish a record for noise as a remembrance to Class OROV IV.

When they marched back, they managed to hold up one another fairly well and they were only two men short when they marched through the gate at midnight minus 30 seconds. Only three guys had lost their caps, but the guards didn't seem to notice.

Next day! Graduation and off!

Back to Eindhoven.

314 was up north again for air-to-air gunnery and according to the maintenance officer, one F-84, the 8T-17, would be ready for flight by 2:15. No test flight needed. It was not a major repair, just parts replacement. They would recycle the gear on the ground a few times and if the pilot would do the same thing once or twice after take-off, just to be sure everything would be fine.

"Why didn't the plane go up with the squadron then?"

"We had to wait for the part all the way from Republic Aircraft in New York,"

"Good thing there is no war on, right?"

"Right!"

At lunch, Shorty joined Frits and asked all about the training in Breda, but he never got any answers, because he would never shut up.

The discussion changed to: "What's next?" and when he heard Frits wouldn't fly till 2 o'clock, he suggested: "How long has it been since you flew a T-6, man?"

"Just about two years!"

"Let's go. Half an hour, forty minutes, maybe!"

"OK!"

Well… .. he regretted that statement the moment the engine started.

He had forgotten the shaking, roaring and coughing of that Pratt-and-Whitney prop engine. He nearly jumped out! Shorty assured him: "When she warms up, she'll be smoother!"

It was, thank God, but the short hop before it took off and the continuing engine noise and vibration kept him on edge all along.

A man got spoiled flying the smooth flying humming jets.

To make it worse, Shorty decided to make some gunnery passes at a freight train and in comparison to jet techniques, he started too low, pulled off too late and generally scared the shit out of Frits.

Shorty took him home in twenty minutes. He laughed himself silly, but Frits insisted: if he couldn't fly jets, forget it. He'd give up flying altogether!

Shorty rubbed that in for thirty more years to come.

The 8T-17 was ready at two.

Frits checked out with OPS and with his luggage stored in the gun bays, he was ready to go.

Man, it felt good being back at the controls of a real airplane.

He remembered his stupid fiasco at Luke after the Christmas lay-off, so he was extra careful with his exterior and cockpit check, going right down the list.

Everything was A-OK. The weather at Eindhoven was overcast, but V.F.R. with plenty of visibility and Leeuwarden reported about the same, just a little more hazy with three miles visibility.

Right after take-off, he lowered and raised the gear twice and informed the safety officer in the tower of that fact.

He received clearance and climbed out through the clouds to 24,000 feet and it wasn't but 20 minutes till he contacted Leeuwarden Q.G.H. for letdown. Leeuwarden confirmed and he was into his letdown through the clouds and would have been on the ground in ten more minutes when Leeuwarden called: "8T-17, this is Leeuwarden Q.G.H., your OPS wants to know name of pilot in command, over!"

"Forrer, Frits, over!"

"Roger Forrer, stand by, over!"

A few minutes pause.

"8T-17, this is Leeuwarden, over!"

"Roger, Leeuwarden, this 8T-17. Go ahead!"

"8T-17, Leeuwarden, orders from 314 OPS, return to base, over!"

"Return? Why? Over!"

"8T-17, return to base, over!"

"Roger, 8T-17, returning to Eindhoven, out!"

"8T-17, Leeuwarden Q.G.H., what is your fuel status? Over!"

"8T-17, fuel fine! Returning to base! Out!"

"Hotdamn.... I should have said I was low and they would have cleared me in unless the weather was socked in. Hotdamn!"

"OK! Back to base."

Back at Eindhoven he called 314 OPS and got Soe-oesman.

"A-are you ou-out of your fu-ucking mi-ind? Who ga-ave you the OK to fly-y and who let you comma-andee-er an ai-aircraft? You haven't flo-own for thee-ee mo-onths a-and you nee-ed to be checked ou-out first!"

"What difference did it make if I landed at Leeuwarden or at Eindhoven. At least you would have had an extra plane that you need and I would have been there too! Who gave the orders? Poublon?"

"No-o, I-I gave that o-order and now I'm gi-iving you ano-other o-order! Get on a trai-ain and get you-our a-ass up here! You-ou he-ear?"

Frits slammed the phone down.

"Ten weeks of not flying requires a checkout? By who's rules?"

He was madder than a pistol. Screw Soe-oesman! I'm staying in town. As a matter of fact, I'm staying in Beppies bed. Maybe for a week. I have some catching up to do.

He did do some catching up, but was on an early train the next morning. It took four hours.

Just thinking about how short a flight it really was, he got steamed up all over again. That worthless ass Soesman, was all he could think about. That worthless ass!

CHAPTER XXII
LEEUWARDEN

"My question remains the same: What's the difference between an instrument letdown here or in Eindhoven? Are the clouds purple or something?"

They were standing in Poublon's office in Leeuwarden.

"Forrer, you should have called here and cleared it, man!"

"But I cleared it with base OPS, the top authority at hand and they gave me the OK!"

"But man, they may not have realized that you hadn't flown for ten weeks and... .."

"No Sir! I told them that! I told them that I had just come in from the OROV IV class and that didn't bother them apparently, cause they gave me the clearance! I made a perfect take-off, perfect flight and perfect landing, so I must have remembered something! But what is the procedure if you haven't flown for a while? Do you get back in a T-33? I haven't seen anyone else do it, so what's different?"

"OK man, Soes, listen: there was no mal-intent on anyone's part. Forrer, you believed that you were properly cleared, Soes, you thought there was an infraction of the rules, so you did what you thought was right, so... we'll drop it!"

Calling him back after Soesman walked out: "Forrer, you took initiative and that's what counts. Besides that, you proved that you don't need a check-out anymore, man, so go fly and have some fun! I'll bet you ten guilders I'll outshoot you man!"

"On the third mission?"

"OK, ten guilders on the third mission."

Back in the crew room, there was no end to the stories that needed catching up and since the weather was socked in, the storytelling moved up to the bar.

Jerry Deerenberg had to show off his new car, the "SPERMATEUR", a Renault Four. It was so small, that the guys had to be fitted in with a shoehorn. When they came up to the mess, Jerry drove towards the wall as if he were going to hit it at 30 MPH, hollering: "Dive brakes, dive brakes; GO!", slamming on the brakes while all four doors opened simultaneously, spilling four hunks as he stopped an inch short. Hilarious!

Later, as the party slowed down long enough for shower and change, Jerry and Frits got into a wrestling match that was utterly unfair. Jerry outweighed Frits by 50 pounds and got Frits carpet burns all over his body but he wasn't really winning and got madder and meaner in his drunken stage, to the point that it took six guys to break it up! The next morning, Jerry apologized all over the place for the previous evening, but nobody wanted to go out with him any more. Too mean when drunk.

Weather remained marginal and with the ceiling at barely 1,000 feet, shooting air-to air was impossible. Flights of two were sent on low-level missions, just to do something!

Frits drew Maarten Schuit as wingman. They had met ten days ago at Matjes funeral and Frits treated Maarten as an old hand. His entire preflight briefing consisted of: "Hang tight! Don't lose me!"

They neatly climbed out to 1,000 feet and once out of sight, they hit the deck, flying between the islands of the North Sea, pulling up to about 300 feet to stay clear of the "afsluitdijk", the dike that transformed the Zuiderzee into the IJsselmeer. After clearing the dike, back on the deck, skimming the waves between fishing boats from Volendam and Marken that were out fishing for eel. At one point, Frits nearly bought it! The sky was gray, the lake looked gray and the baskets on six foot poles on top of the dikes were certainly gray. Originally they were painted Red, White and Blue, but by now they were weathered and gray! Wherever the dike made a sharp turn, a pole and basket were posted as markers for the boat traffic, not for low flying aircraft!

At 420 MPH, they skimmed the water till they come upon the dike along Friesland and had to pull up a little to clear the 12 foot dike. Unexpectedly, there was a pole and basket right there and instinctively, Frits raised his right wing in order to clear the obstacle and never knew by how many inches the left wing cleared the dike! He missed the basket, but had his left wing hit the dike, it would have been sheared off and the plane would have cart wheeled and torn itself into smittereens, with little pieces of Frits being blown all over the Frisian polder!

Maarten never noticed! He was too busy hanging in there!

All of a sudden: *"FANCIER BLUE, BOGIE, 10 O'CLOCK! TIGHTEN UP! SIGHTS ON!"*

Maarten tucked in closer and together they made a pass at a lone F-84 that turned out to be a Dane. *"TATATATATATATA-TA"*, they simulated a gunnery pass at the lone bird and passed underneath him. The Dane never saw them! Blue flight reversed their turn and made a pass from the other side: *"TATATATATATATATATA!"* and the Dane remained oblivious!

"Fancier Blue, join on his right wing. I'll join on his left. Watch out for a sudden turn into you! He may get startled!"

They were back on his left side, so Maarten slipped under and up and when the startled Dane looked out, he had inherited two Dutch wingmen. After waving at the dumb ass, the guys dropped back, gave it the gun and in tight formation flashed in front of the surprised F-84 dude. "That'll teach him to fly over Holland!"

Back on the ground: "Holy shit! You hadn't been checked out in the squadron yet? Oh, shit!"

"This was my first flight here!"

"Well, welcome to 314. How do like the way we indoctrinate our new people?"

"Frits, do you want to know the truth?"

"That would be a novel approach!"

"You worked my ass off up there, scared the shit out of me a few times, but man,,,,, what a way to fly!"

"I guess that was a compliment!"

A notice from the Navy had been posted to "All Pilots." The Navy wanted pilots to fly in New Guinea.

They offered: increase in rank, one level! Double time towards retirement for serving in the tropics.

350 flying days a year.

Flying off Holland's lone carrier: "The Karel Doorman".

One drawback! Flying props again! The five-bladed Sea-Fury!

A lot of the boys who had little hope for a commission saw this as a chance to get one.

Guys who were tired of just 100 flying days a year in Holland saw this as a chance to get some real flying in.

Others thought about speeding up retirement and still others saw it as a chance to see another part of the world.

After a lot of beers, a fairly large group submitted their applications through channels, Frits included, and waited for answers from the Navy.

They did not get that far!

All applicants were summoned to "POOP's" office!

"ATTENTION"

While tearing up the applications in front of their eyes, he said, dramatically slow: ***"NOBODY LEAVES MY SQUADRON! UNDERSTAND?"***

"DISMISSED!"

"Well, I didn't wanna fly props anyway!"

"Who the hell wants to be in the Navy?" They consoled themselves somewhat and continued with their task of having fun!

This years early spring never came. It had not been a good exercise for Air-to-Air at all. Usually, cold freezing weather prevailed at this time of year, but not this time. Fog, low clouds, drizzle and sleet was a near daily experience, so they lumbered back to Eindhoven for routine flying and living.

Frits got to fly element leader frequently with different results. One day, with Katjang (peanut in Malayan) close on his wing, they cleared the ground and when out of sight and radar range, they hit the deck. Just south of the Belgian border was an auxiliary landing slab that the Germans put down during World War II, but was utterly unused and they hit it doing 500 MPH. Frits low over the runway, Katjang, slightly behind, over the parallel taxi strip. Like Arizona all over again! Right on the deck! A fantastic sensation of speed and power with the concrete passing just a few feet below them.

Adrenaline ville!!!! WOW!!!!

Up to 1,000 feet over land and down to sea level again over the blue waters of Zeeland! Then: Pitch-up! Up, Up, Up! 10,000 feet over Rotterdam in a slow wide barrel roll, upside down over the big harbor city and continuing the wide circle with Scheveningen, Kees Rog and The Hague in the center of their diving roll.

Back to 500MPH and back on the deck skimming the 15 foot waves of the North Sea 2 miles off the coast of Scheveningen. Looking back at his wingman, he startled! Frits thought he was barely skimming the waves and there was Katjang, flying 2 to 3 feet lower!!!

"DAMN IT! STACK IT UP!" No name, no number. Nobody would know but the wingman. That S.O.B. has the balls to stack down, to actually fly lower than his leader!!!

On the ground he answered: "Well, you were showing off how tough you were and I wanted to show you that I can be tough too!"

"Don't ever stack down again on the deck!"

"OK!"

Frits was sure he didn't mean it.

On a flight of four at 18,000 feet, Benninga leading, Frits #3, all of a sudden: "Blue lead, this is Blue three, overheat warning light on! Over!" The red light, high on the instrument panel glared brilliantly at him. He pressed it, as to test it, but it stayed on.

"Blue 2 and 4, break off and continue flight. Blue 3, reduce power, maintain heading. I'm dropping down to take a look. Drop dive brakes!"

"Roger lead, dropping dive brakes, tailpipe temperature 610 degrees, no vibration, over!"

#3, Blue lead, no visual flames!" Henk was sitting slightly below and behind Frits' tail. "If in doubt, #3, stopcock engine, trim for glide to the IJsselmeer and bail out. Over!"

"Roger, this is three. All instruments are in the green, tail-pipe temp normal, no fire warning light, no thrown buckets, no vibration, hydraulic pressure normal, light will not go out when pressed, over!"

Katjang

"Three, this Blue lead, do you want to take it in?"

"Roger, this is three, returning to base."

"Roger three. Don't hesitate. When in doubt, bail out! OK? Good luck!"

Roger, three out!"

With reduced power he eased his way down and over to the base, where he made lazy circles in order to burn out his tip tanks.

All the instruments indicated normal engine operations so he just kinda sat at 5,000 with too much time to think.

""Is this what I want to do the rest of my life? Taking chances with these jets? Look at Gilliam, Baetens, Schipper, Rog and Matje Willems? Jeez, it could happen to me sometime. I think I'm hot shit, but so did they! Matje and Schipper were a lot more experienced than I and see what happened to them! Maybe one day I might want a family and kids! Maybe I should think of doing something else!"

For the first time, after flying for nearly three years, doubts were settling in.

"But what else would I do? Being a career Officer is not a bad future! What else could I do? Commercial? No way! It's this or nothing!"

The possibility that the plane could blow up any minute was remote, even though a red light was staring him in the face, telling him that it was about to happen.

It was getting to him.

He didn't dare loosing sight of the runway and kept checking and rechecking all the indicators and instruments in front of him, ready to bail out at the first drastic change.

By the time he came in for landing he had convinced himself he would resign as soon as his three year contract was up in a couple of months. He had had it! That was it!!!

Of course when the compliments started coming in about his cool handling of a potential disaster, his confidence grew by leaps and bounds. After filing his report, it was found that a "nine-cent fuse" was at fault and a lot of people might have jumped out of a million dollar plane because of that! A pat or two on the back did wonders for his morale.

Then another blow!

Geppy Steenhuis had gotten killed at Volkel!

"How? How?" "He was a big-mouth, but he was a good pilot, so what happened to him?"

"Unofficially... . A mid-air collision at 1,200 feet in the traffic pattern. Gebby ejected, but ejected too slowly and the vertical stabilizer caught up with him and knocked him cold, so he never got out of his seat and got killed on impact when he hit the ground."

"He ejected too slowly? How can you eject too slowly? The rockets shoot you the hell out of there, right?"

"Oh yeah, the guiding tubes along which the seat blows upward were bent slightly and slowed him down!"

"I can't believe this! Bent tubes? Are we all flying around with BENT TUBES? God forbid!"

"NO, NO, NO! I'm not going to another damn funeral and that's final!"

Frits was adamant! He was merely being asked, since he was in the OROV class with the guy, but his reaction was as if he was disobeying an order!

"I don't care if the Queen dies! I'm through with funerals for life! They're idiotic, painful for the families and nothing but a hypocritical showoff! If you're going to show respect for someone, do it while he's alive! Not when he's dead!"

"My, my, you're getting short tempered!" Katjang was playing house psychiatrist.

"Well, that's how I feel. I'm not gonna die in an airplane, but I'm gonna write a will stating that I want a party thrown when I die, with people toasting all the good things about me in life!"

"When are you gonna start?" Katjang is still goading him.

"Start what?"

"Do something good!'

"Screw you, you bastard!" He broke up laughing.

"Let's start now! Let's get a beer!"

All the N.C.O. pilots were now in Camp Zeelst, where Leo Hermkens tended bar.

Leo was a typical southern man with a great sense of humor but an atypical complexion and hair. He could have been Greek. Olive tan skin, black eyes and hair and a perpetual smile. He loved bartending for the young guys and did everything in his power to accommodate them. That ranged from frying eggs to drying drunken tears. A few beers usually tended to brighten their smiles again.

Time was flying by! Literally! Weather improved and more hours were put in on the stick. That always helped morale.

The base got caught up in a NATO exercise that restricted everyone to the base for 72 hours.

The pilots slept in the crew room and were on standby 24 hours a day.

A flight of four sat at the end of the runway, plugged into their A.P.U.s (Auxiliary Power Units), ready to roll in seconds.

A lot of times they were scrambled to intercept "bogies" somewhere over western Europe and chased other aircraft all over the sky. Sometimes, they'd climb 45,000 foot, barely flying anymore, just sit and watch British "Canberras" move overhead at 50,000 feet and not being able to do a damn thing about it.

Other times, they'd sit at the end of the runway for hours, just to be replaced by another team of pilots without having flown at all.

The two squadrons, 314 and 315 rotated duties.

One night 314 would handle the standby and 315 the next.

When on duty, the guys sat in their planes or in the crew room with red lights and sucking 100% oxygen out of their portable containers.

Night vision improved remarkably when the blood in the eyes contained pure oxygen.

Yet, in spite of that, a RECCE pilot from the 316 squadron never gained altitude after take-off and flew into the living room of a farmhouse, a few miles from the base, killing himself and six people at the dinner table.

On the night that the squadron was off stand-by, it was time to party!

Because of Holland's affiliation with Indonesia, "RIJST-TAFEL" (rice-table) was a favorite Dutch dish and some of the old"East Indie Pilots" were excellent gourmets.

So, on the "off" night, at 5pm the beer would start flowing and the night's festivities began.

Card games and chess were popular quiet games. Arm wrestling and leg wrestling were popular loud games. When some people were busy cooking the SATE and RICE, the others would make wild bets and did some beer inspired hooting and hollering at the contestants.

The toughest test was: "knee-wresting" and the champ was Karel ten Hove. The guy was a real bodybuilder and was unbelievably strong. He could kill someone with his bare hands. With his knees, he nearly did.

Knee-wrestling consisted of someone sitting on a table, his legs around the waist of the victim standing in front of him, legs locked at the ankles. The job now was to squeeze the victim till he screamed or for the victim to tighten his stomach muscles so he could withstand the pressure. It was a hell of a game, but no one had enough leg strength to make Karel scream, but he put so much pressure on one guy's kidney that he fainted and the paramedics had to be called out to take him to the base hospital. Nice games!

All in all, they would have a blast! All night long, eat, drink, sing, play games, tell jokes and laugh! Old fashioned good fun.

Usually by midnight, things died down and everyone zonked out on their air mattresses.

One night, something went haywire and 314 was scrambled on their "off" night. Frantic phone calls by "Poop" didn't change a thing. 315 was already up and flying and 314 had to scramble all available aircraft. Some guys were so loaded, they couldn't even wake them, but black coffee and 100% oxygen got 12 F-84's airborne in twenty minutes and back down one hour and ten minutes later. To everyone's surprise, there were no mishaps and no lost planes, so everyone got back in their sacks.

Imagine the looks on their faces, when at 7 in the morning coffee and breakfast started smelling up the crew room and Jan van den Engel started a story: "I had the most realistic dream! I dreamed that somehow we got scrambled and twelve of us got to fly a mission and went back to bed!"

When subtle laughing erupted he became indignant: "Yeah, man! It was so real that I could swear it happened! You, Poop were leading, man and you were in it too and you and you... !" until laughter drowned him out!

Another bombshell... . Ger Walhout moved out of Kamp Zeelst. He had gotten his promotion to Warrant Offer and moved into the B.O.Q.'s.

Frits ran out to check his mail... ..nothing...

The next day: nothing... .. and the next... . and the next... Nothing!

Not a word! From nobody!

After a few days he cornered the administrator, a Sergeant-Major and a hell of a nice man.

"You don't know? Nobody told you?"

"What am I supposed to know?"

"You've been placed on a three months probation, holding up your promotion!"

"Whose orders? Poublon or Breda?"

"Neither, it was a mutual decision!"

"Those bastards! Those rotten bastards! And they didn't have the balls or the decency to confront me and tell me about it? Those lousy cowards! Now you understand what Joop Jonkers meant when he refused to go to OROV VI. He told them: "If you're typical of what officers are to be like, I don't want to be one!

He was right, you know! Screw them! It's kinda disgusting though, that Poop nor anyone else would call me in to talk about it. Well, that helps me decide. Thanks, Sergeant-Major!"

Even months later, neither Frits nor Poop had brought it up and now he had to make a decision.

Up to now, reserve flying was done at the squadron. Men leaving the service had to serve 4 days a month and fly 4 hours a month in order to stay "qualified." This was for 6 years! Most guys loved it! They were paid for the 4 days, for their flight time, plus, their monthly flight pay of 125 guilders. For 4 days work, that was great and the schedule was flexible enough so they could fit in their flying days when they wanted. Most of them became weekend jet-jockeys and loved it.

This was going to change!

Reserve flying was going to be done on Piper-Cubs! No more jets!

All this was because of the advent of the F-84-F, the "THUNDERSTREAK." The F-84-F had a swept-back wing and would replace the old straight wing F-84-G's. It could go through the sound barrier, where the old one could only go Mach .83, or 623 MPH at sea level. That was an appealing thought! Going through the sound barrier! Making like Chuck Yeager! The "Streak" was heavier but much more powerful and more maneuverable, all good reasons to want to fly it and everyone was looking forward to getting checked out on it.

One big catch! "If you don't sign up for an additional tour of duty, you don't get to fly the "F"." "You go to Pipers!"

So if Frits quit, that would mean the end of his jet-fighter flying and that was a hard decision to make.

Another consideration: Unless Poop gave him some assurance that he would be promoted after three months, what was the point of signing up for an extended tour?

He waited to see if Poop would make any moves at all. Guys kept leaving the squadron in droves, but by the same token, new guys kept pouring out of America, but it seemed like a different breed.

The "new" breed had signed for 6 years AFTER getting their wings, so a total of eight years. WOW! Very serious guys with very serious intentions! Frits' generation had signed for three years including training, so they saw it all as a great adventure, nothing serious about it and most of them did not have a serious bone in their body.

Maybe Poop was opting for the more serious brand. Screw him!

Ike quit and got married.
Keesje Bos went to fly in Iran.
Jan Willem Mooy went to Swiss-Air.
Maarten Schuit left to further his studies.
Joop Jonkers just left.
Jerry Deerenberg resigned as of September One.
Jan Mante quit and got married. Frits was his best man and had a ball!
Henk Beninga was joining Swiss-Air.
The picture was changing and Frits hung on, undecided.

A few things helped him decide, though!

For the International Air Show that was to be held in Holland, the Italian F-84 stunt team was temporarily stationed in Eindhoven and that added some excitement. Their planes were painted multi-colored, but the most interesting thing was their maintenance plane. It was an old Savoia-Marchetti. It looked like a JUNKERS JU 52 of corrugated aluminum, just like the old FORD TRI-MOTOR. Going up in that antique was as fascinating as scary. The damn thing creaked and groaned, it was drafty and cold, but it flew like a duck. A big duck! It shook and rattled like hell and sometimes it seemed like it was flapping its wings in order to stay aloft. What a machine!

On the day of the big show, Eindhoven would do a big fly-by with 32 planes and Frits would fly standby. "Standby" was sitting a few thousand feet above and away from the formation and ready to assume any position if someone fell out for mechanical reasons.

Being that he was going to be the last one up, he sat in his cockpit, patiently watching the planes take off in two's, one pair after the other in rapid succession. Poop in the lead plane had started a wide circle allowing the pairs to join up with him in box formations. The ultimate result would be: Four boxes of four by 314, followed by four boxes of four by 315. It should look great!

Poop and his wingman had been the first ones off and Soes

Jan Mante and bride

with Willy Wishaupt on his right went next.

Poop had started his wide left-hand circle. Soes was to slide in Poop #3 position and Lt. Wishaupt underneath in the #4 position.

Hans de Jong, 314's dashing Latin, headed up flight #2. His

element started rolling as Soes left the ground. By flying a slow climbing turn, Poop gave everyone a chance to cut him off and get in position. For some reason, maybe he lost Poop out of his sight, Soes stayed low with Willy on his wing. By staying low, he picked up more speed than the others, but when he went to gain altitude, he started to fall behind Poop, steepened his bank and then came up, nearly standing on his wing below Hans de Jong's formation. Hans obviously couldn't see through the bottom of his plane and never knew what hit him. Just as his #3 man screamed: "Hans, pull up!" Willy's right wing hit underneath Hans' left wing, rolled underneath Hans and hit his right wing and elevator. Willy's F-84 broke apart and hit the ground in seconds in a big fireball explosion!

Hans, too low to bail out, rode the crippled plane down in a perfect forced landing in a marsh.

Soes never saw what happened.

Poop assessed the commotion on the radio and calmly had Soes join up with him. He told #3 of Hans' formation to join on him in the #4 position and told #2 and #4 of Hans' group to pull up and join the reserve group that was taking off last. He ordered 315 to pull in the slot of the 314 formation and his #3 an 4 flights to assume positions on his left and right wing. Business as usual, cool as a cucumber.

Now the formation consisted of 28 F-84's in seven boxes of four.

Three in line, with two boxes on each side. Not a bad looking formation.

Frits, now with two wingmen, sat on top looking down on the show and the parade of aircraft flying over the field near Rotterdam. It was pretty!

Planes of all types, American, British, Swedish, French, Italian, Danish, Norwegian, Belgian and Dutch, one flyby after the other.

Too bad they had to make it back home before fuel ran out.

Back home a lot of controversy sprung up. Willy was as dead as a doornail, no doubt about that. Hans was OK! It was a miracle, that the plane could fly at all, let alone, make a perfect wheels-up landing!

The question: *WHAT CAUSED IT?*" made Frits boiling mad. *"IT'S NOT: WHAT CAUSED IT? IT'S: WHO CAUSED IT?"*

"Soes should be hung for murder!" was his official opinion and he was sure that the tower operators would concur.

At the debriefing, he drew where every element was at the time of impact and every element was in it's proper place except Soesman!

There was no more flying that day but also for another reason. That evening, a party boat with all the show participants would sail the Rotterdam harbor from 7 to 12 with all the food, drink and music a man could want.

The guys changed clothes, teamed up and headed for Rotterdam and party time! It was a great boat, with bands and buffets at several levels and the food and drink was the finest imaginable.

There were few women aboard, but he consoled himself with the thought that he had to end up in Beppie's bed anyway, so he made the most of it.

If he hadn't run into his old roommate "Flory" from Kinston, it would have been a super night. Flory had a very dampening effect on him. Not that it wasn't good to see him again and exchange experience, but comparing notes became depressing.

Comparing careers, income and achievements was a total letdown!

Flory had flown F-86's in Korea and was now a First Lieutenant. Frits was still a Sergeant. Flory flew F-100's in Germany, Frits was still in the old F-84-G that was half as fast as an F-100.

Flory earned $1,080 a month, Frits about $180. Flory drove a little red Mercedes convertible, (used) Frits rode a bike.(used)

It was just demoralizing!

"Why couldn't I be in the American Air Force, fly modern planes, earn decent money, rapid advancement? Why? Why?"

When in the next few days Soes continued as OPS officer and Willy was buried in the Old Saint Servaas Cathedral in Maastricht on the day and the hour that he was to have been married in the same church, Frits reached his decision. He was leaving!

He didn't know what he would do, he didn't have a vocation, he didn't want to be a member of a large, structured organization like an Airline like KLM, or any other big company. Too many bosses! The first jet-liners were due out in the next

Ready to go

few years and word was that every airline wanted men with jet experience, although it was rumored that KLM preferred pilots with "multi-engine" experience, such as Gloster-Meteor jocks.

"Well, no matter! No straight and level flying for me. Maybe my own business? Maybe a flight instructor in Taiwan?

He rally didn't know what to do and the economy in Holland wasn't particularly rosy, so that didn't help either.

Sometimes he wondered what he would do if his promotion came through and he was assigned to fly the new

F-84-F?

He was too proud to approach Poublon and Poublon did nothing, so the standoff continued!

CHAPTER XXIII
GOODBYE 314

The parties didn't end, they seemed to multiply!

Ike had the whole gang over to his father's house in Deventer.

Red Verdonk, Karel ten Hove, Karel van Veen, Harm Appelo, Arie de Jong, Katjang and many others.

Sleeping available for all! No drunken driving! Great Idea! Great house and a great warehouse! Frits had never seen so much beer under one roof. It was stacked from floor to ceiling along the both sides of this huge warehouse in rows of cases upon cases and kegs upon kegs!

Ike's father, (a character just like his son)told him: "You look like you're seeing the gates of heaven open up!"

"I do! I did! I sure do! This is beautiful! What a place to die!"

Nobody died, but the old folks Eijkelenkamp nearly died laughing at all of the antics, songs and jokes that they had heard about all those years, but never experienced. There was no limit to the beer flow obviously and when Old Eijkelenkamp brought out some good Cuban cigars, even Frits lit one up, but later on could not understand why it didn't make him sick.

The important thing was; it didn't!

Somewhere around three in the morning, Old Ike woke up Frits: "Come help, we got a problem!"

The problem turned out to be Karel ten Hove. He was sitting in his V.W. in the warehouse, threatening to run down the overhead door, if Old Ike wouldn't open it!

Frits tried to convince Karel to go back to sleep, but Karel insisted: "I'm leaving immediately for my immediate friend Agnes and if that door isn't opened immediately, I will immediately crash it!"

"But where is Agnes?"

"Agnes is immediately in The Hague and that's where I am immediately heading because I'm immediately in love and I need to immediately see her!"

He was drunk as a skunk!

"But Agnes is a Wave and she's in the barracks and she can't come out until seven at the earliest! It's three now, so go to sleep a few more hours and you'll be in time for breakfast, so turn off the engine.."

"Get your fucking hand out of here!" Frits had reached for the key, but Karel was too quick and he started rolling up the window.

While leaning on the horn, he rolled the window down again and hollered: "I'm crashing the door now! Immediately!"

He raced the motor, moved about ten yards, but stopped within inches of the door.

Mr. Ike looked at Frits who shrugged his shoulders and Old Ike walked over, turned a switch and opened the door! "Don't kill yourself Karel!"

"Immediately!" and he was gone.

Turning to Frits, old Ike said: "What could I do?"

"Nothing! Let's go to sleep!"

"You know Frits, you guys are a crazy bunch!"

"Immediately, Mr. Ike, immediately!"

Arms around Frits' shoulders, a smiling Mr. Ike walked him back up the stairs. "What a bunch! What a bunch!"

A much more sober lunch was served up by Mrs. Ike, who, by coincidence, had gone to school with Frits' Mam in their hometown of Winterswijk. Because of that, she had a soft spot in her heart for him.

With all the guys going in different directions in life, there was a strong sense of camaraderie amongst them as if they could feel in their bones that their friendships would not evaporate when their uniforms were stripped off.

It was about the last time that all the buddies broke bread together and it left them both melancholy and excited.

Melancholy, because it was an end of an era for them. Excited, because of what lay ahead of them: "a lifetime of unknowns."

Another highlight of the season was Red's wedding to a dark haired beauty, called Toni.

Red had gotten a lot of ribbing from the guys because of the way she tamed that red haired stallion and now they had made their decisions. Red would make the Air Force a career and tie the knot in matrimony.

With Frits as best man, it should be a successful start to a successful life!

At least, that was Frits' opinion!

The wedding was held in a beautiful medieval castle that now served as a city hall. A radiant black haired bride, accompanied by a dashing red topped groom in his new Officers uniform with two wings adorning his chest: the gold and orange Dutch "vink" and the silver American pilots wings.

It was a beautiful ceremony in a beautiful setting, with two beautiful people pledging beautiful promises to one another.

The reception/dinner was definitely the highlight of the event!

Red had invited the whole squadron, past and present.

Always ready for a party, the pilots flocked in by the dozens and before long, they were the center of attraction. Singing some of the most distinctive songs, dirty or otherwise, they let Toni's family know what she married into: "A live, active, hilarious crowd of semi-intelligent madmen!"

"Gaston", a relative newcomer from south Limburg, married to a hillbilly from Tennessee, demonstrated "goldfish" drinking in his beer and nearly got arrested. Kareltje had tried to rescue the goldfish by sticking his finger way down Gastons throat and got him to throw up violantly. The fish died anyway! Detectives showed up to investigate a complaint of "animal cruelty".

He got off with a warning.

At least Red and Toni went off on their honeymoon without any scandal or embarrassment.

Frits did have a last major run-in with Soe-oesman somewhere deep in Germany.

The squadron was off on maneuvers once more, this time on an R.A.F. base in the British sector, teamed up with the Brits against America, France and Italy.

War games were supposed to be primarily for training purposes, but they also could be good old fashioned competition.

Living in tents always added a boy scout kind of atmosphere and lent a sense of "roughing" and "adventure" to it!

This time they were at a "Canberra-bomber" base, called Güttersloh, camped out amongst the pine trees.

The weather was at its best, cloudless and warm.

The flying was great, although there were very few serviceable aircraft. No new F-84s of the old type were being added and they kept breaking down more often and parts were hard to come by. So, the pilots weren't as busy as the airplanes, which allowed time for sunning and relaxing. On one of the return trips from the flight line in an open truck, Soesman decided to run his mouth again. At the drop-off point for the N.C.O.'s he hollered: "Tomorrow mo-orning, 6:55 sharp, the tru-uck will leave here and at se-even sharp from the o-officers a-area. Anyone, not on time will no-ot get brea-eakfast! Six fi-ifty five, remember!"

The truck with the N.C.O.'s pulled out on time and pulled into the officer's area at 7am.

At 7:05, Frits ordered the driver: "Take off!"

"But Sir... "

"Take off, that's an order!"

Well, the officers got a very late breakfast and Soes ordered Frits to dig latrines.

"Good God, it's tempting to tell him to shove it, but I only have two more months to go on active duty, so how much of it can I still screw up?"

His mind was racing, but he said: "Where do I find a shovel and how big do you want it?"

A few hours later he returned. "Thanks, Lieutenant, I needed a tan and a good workout, I'm getting too fat. Besides that, it gave me time to think. I've finally finished my new squadron song. Do you wanna hear it?"

"Fu-uck no! Ge-et outta here!"

"Don't get ulcers! Give'em!" he said to himself.

The other guys agreed that digging latrines was degrading punishment for a pilot, but Frits philosophized: "At least I don't have to fly with him, so I'm still alive! You suckers may not have that choice!" So much for that old 314 spirit!

"But guys, I finished my 314 song. Wanna hear?"

"Alright! Burn our ears!'

"It's to the tune of "JUNIOR BIRDMAN." You 'member?

"OK, spill it!"

Into the air, Three-One-Four men!
Into the sky, all th'world 'round.
Into the air, Three-One-Four men,
Get your gear off from the ground (off from the ground)
And when you see the enemy flying
And you shoot some down once more,
You can be sure, that they'll be crying:
"Must be a man from Three-One-Four!"

Into a bar, Three-One-Four men,
Order a drink, it's time for beer!
Into a bar, Three-One-Four Men,
Fill the joint with joy and cheer! (with joy and cheer)
And when you hear the voices ringing
And your glass is filled once more,
You can be sure, they'll soon be singing:
"The song of famous Three-One-Four!"

Out with a girl, Three-One-Four men,
Out on a date, let's have fun!
Out with a girl, Three-One-Four men,
Show her the way it should be done! (it should be done)
And when you have your baby sighing
And she cries for: "more, more, more!"
She sure will say: "no use denying,
"Must be a man from Three-One-Four!"

"Wow! Hoorah! Write that down, man!"

"I think it stinks!" that was Jan van den Engel.

"You oughta smell yourself sometime!"

"Watch how you talk to an officer!" He was a Second Lieutenant by then.

"Smell how you smell!"

The new pilots may have been great guys, but that kinda adventurous pirate kinda spirit of the "three-year guys" was definitely missing.

Tearing through the skies with a new man on his wing, Frits didn't find the fighting spirit of a Katjang, Ike or Maarten. As a matter of fact, he had to modify his attitude toward new people.

A more defined briefing, showing #2 what to do, where to position himself and to hang in there no matter what!

What brought that up was a couple of real scares.

Up one day with Jan-Tieme on his right wing, they were playing their usual game of low-level passes and high altitude antics, when they came upon a German lake, cluttered with sailboats.

What a target! Flying about level with the decks of the sailboats, they ravaged them with their jet blasts, blew them over and made them dance on the waves! "That ougta teach'em to start a war!"

Looking back over his shoulder, Jan-Tieme was sitting up so high, he couldn't possibly do any damage! That crummy chicken!

He said to himself: "Well, Holland will have to fight the next war with clowns like that! More power to them!"

On the trip back, they climbed to 10,000 feet and ended up above a near solid cloud deck. No problem, except: they'd have to go on instruments for their letdown and approach, but that was too dull to be considered! Somewhere over the eastern part of Holland a hole appeared in the clouds and it didn't take a second to reach a decision. "Tighten it up #2" and down he went in a tight spiral as if he was making a vertical barrel-roll toward the ground. After about three corkscrew kinda turns, he was in the clear below the clouds and looked for #2. Not on his right wing where he was supposed to be and not on his left either!

"#2, this is Fancier Blue, what's your position? Over!"

"Fancier Blue, this is Blue 2 at 10,000feet over the overcast! Over!"

"That sonofabitch! That bastard! Chickened out and didn't follow me down! Hotdamn!"

Now he was faced with the fact that he had a new man sitting up there, possibly panicky and ready to go on "recorded" frequency in order to request an instrument letdown. "Hotdamn! How do you explain losing a man?"

"Number two, home in on "Lobith", start left turn at angels twelve (12,000 feet) and wait for join-up, over!"

"Roger, left orbit at angels 12 over Lobith, out!"

At least if anyone was listening in they would not gather from that conversation that the flight leader was at 5,000 feet and the wingman at 12,000.

Now he had to make a slow, spiral climb through the same hole.

It was a tedious job, because any minute he might find himself in the clouds, making an unauthorized instrument flight in the soup.

Finally, he found #2 in his left orbit and got him back on his wing. This time he did what he was supposed to. Made a letdown and approach on radio and instruments. Dullsville, USA! Boring stuff!

No wonder the guy didn't drink. He had no fighter spirit whatsoever! Maybe Holland would get bombers one day and the guy might have a future!

On the ground, he scolded his #2 man more lightly than he would have liked to, because he wouldn't put it past him to go crying to his superiors about the rough treatment he got!

On another occasion a flight of Danes was visiting the squadron with the same restrictions they had in Denmark. "No four man formations, no aerobatics, no flying after drinking, and so on1"

As a result, the Danes flew sole #3 positions with a Dutch flight of two.

On an overcast day, Frits was up with Jan-Willem as #2 and the Dane as #3. The clouds hung at about 2,500 foot, but the visibility was excellent, so they set out on a modest low-level flight at 500 foot over Germany. When they crossed the Rhine near Cologne, he made sure they were clear of al high-tension wires. He guided his trio up the Ruhr valley, constantly banking right or left, following the river. That Ruhr valley is one of the most beautiful spots in the world, whether walking, biking, driving or flying through it, so Frits was immensely enjoying himself. The only danger really would be, if the clouds lowered in the valley and the mountaintops were obscured, they'd have a hard time turning around. Other than that, the flight was nothing but a beautiful joyride!

Having found a spot where they could safely make a U-turn without pulling into the clouds or crashing into the mountainside, they turned and started their flight westward toward home base.

#2 and #3, the Dane, were in fine position at that point and Frits happily started winding his way back. The turns in the river, sometimes sharp turns, demanded all his attention and not until they had leveled out west of the Rhine, did he have a chance to look back. #2? Doing fine! #3? "Where the hell is #3?"

A tight turn to his right, back to his left... .. no #3 man! No one in sight! "Hotdamn!" Where could he be? Splashed all over a mountainside? God forbid! How would he explain this back home?

"Up in the clouds? If he's not too disoriented to begin with, he's certainly up there without a clearance, maybe creating a mid-air collision."

"Hotdamn! Dumb Dane! And right near HAHN A.F.B.! Right near Cologne International Airport! He could be wreaking havoc up there!

"Fancier Blue Three, what's your position?" He was sure Hahn could hear him! Embarrassing! He could just hear them saying in the tower: "Some Dutch Jackass lost his wingman!"

No answer! Nothing!

Not enough fuel to go down the Ruhr again and even if he did, what should he look for? Airplane parts in the woods?

"#2, change channels!"

"Fancier Blue Three, what's your position, over?"

Nothing. "Better go home! "#2, back to fighter channel."

He's talking to himself: "Let's hope the jerk can find his way back to Eindhoven."

Once over Holland again, some 90 miles out, they started a slow circle at 2,000 feet, calling and calling again!

Finally... .. "Fancier blue, dis is tree!"

"Halleluia!"

"Three, what's your position?"

He was over the Maas, (the river Meuse), but he didn't know exactly where.

"Fly north along the river at 1,500 feet. Call when you see a town with a lot of bridges, over!'

"OK."

When he called in over a town with a lot of bridges, the town straddling the river on both sides, Frits knew he was over Liege.

Ten minutes later they spotted him at ten o'clock and slightly below. "Hang in tight, #2!" A very tight turn got them right on his tail and a minute later they were back in formation, heading home.

"WHEW!" "Talk about relieved!"

"I just don't see you any more," The Dane told them over a beer and that was all he had to say for himself! It did cost him a case of beer!

Flying with his known and trusted buddies was a lot more fun.

At the OROV classes in Breda, the constant bragging between the F-84 and Gloster-Meteor jocks was always enhanced when planes of one or the other type came buzzing over. It brought risks with it, because being reported for being too low or doing something acrobatic could result in punishment of all sorts. So showing off was always at least "moderate."

Red and Frits connived to join up for a four ship formation flyby that would not get them caught.

When both of them would be up at the same time with a flight of two, they'd join formations by racing up and down the right side of the Rhine river at 1,000 feet until they'd meet and then form a flight of four. With Red in the lead, Frits flew his box and the wingmen took up the left and right positions.

Having good wingmen, they could tighten up the formation with the wings overlapping as if they were the "Thunderbirds".

Their passes over the Academy gave the F-84 pilots on the ground bragging rights for at least 3 more weeks. Frits was tucked in so close, that he looked right in Red's tailpipe and the exhaust burned his stabilizer black with soot.

It was a wonder that they never got caught. Either they were never reported or nobody suspected innocent flights of "two" to be the perpetrators.

For whatever reason, they kept getting away with it and of course would always toast their successes.

The boys attending the OROV classes, would later on relay their enthusiastic endorsements and buy more beer!

On another occasion they had themselves a blast.

A sort of "*search and destroy*" mission was on the schedule in conjunction with an Army unit on maneuvers.

"Intelligence" had pinpointed a division of armored vehicles near a bridge crossing, somewhere in the middle of Holland. The area was heavily wooded and hilly, intersected by a canal as straight as an arrow. The armor was suspected of trying to cross the bridge under the protection of darkness, so they had to be "destroyed" during daylight hours.

Henk was leading the flight and Frits was his element leader. On their first pass, there was no sign of an enemy anywhere! Not a soldier, not a truck, not a tank! Nothing! Several more passes with the same result!

Henk called off the attack and lead his flight away as if they had given up on ever finding the target.

Fifteen minutes later, though, they approached the target again on the deck, hitting the canal about 5 miles south of the bridge. A prop plane seems to send it's engine noise ahead of itself, but a jetfighter outruns is noise, so Henk settled in between the canal banks with his #2 man off to his right and behind and Frits with his wingman trailing by about half a mile

The bridge up ahead was between the planes and the division, as Henk had figured and that kept the jets out of sight until it was too late! Within seconds from the bridge, Henk pulled up about 50 feet, barely clearing the bridge and hollered:" There they are! Fire at will!"

It was a sight to behold! Hundreds of soldiers, spread out on the left bank sunning themselves, dozens of them splashing in the water and clothes and towels everywhere! "Total surprise!" The moment Henk and his man cleared the bridge they nosed down toward the water: "Dive Brakes, dive brakes: GO! Full power!"

The scramble was unbelievable! Frits had pulled up at Henk's first call and had a 50 yardline kinda view of the spectacle! The soldiers scrambled up the banks and out of the water, camouflage nets were being pulled back over the exposed tanks and trucks between the trees and to make it worse, some ran back and hit the water!

Dive brakes interrupt the smooth flow of air underneat the plane and causes a loud rumble! The jet engine, at 100% power delivers a deafening roar, that would scare the piss out of an elephant! Imagine what it might scare out of the poor soldiers! The bathers were stretched out over quarter of a mile and when

Henk and his wingman pulled straight up when over the crowd, their jetblast blew soldiers over as they ran and scattered clothes and towels all over the landscape! Frits hung a little more to his left, looking tank commanders and truck drivers straight in the eye before pulling up at the last possible second, sending his jetblast right down to his victims!.

Henk had made a steep wingover, turning 270 degrees and as soon as Frits pulled off, he hit again on the Army from across the canal. Frits followed suit from the opposite directions, thoroughly enjoying himself, mouthing:"TATATATATATATATATA", smack over the scrambling troops on the bank! Napalm would have cooked two hundred of them, he was sure! On the third pass the troops had mostly disappeared into the woods, but enough tanks and trucks were still visible to make for a good bombing run, had this been real.

Thoroughly pleased with themselves, they retreated for home before gas would become a problem. This was the life of the jet fighter pilot he loved the most!

One day, though, he nearly "bought the farm", as Chuck Yeager would have said.

A parade was scheduled over the palace of Queen Juliana at Soesterberg and both the *tactical,* as well as the *air-defense* bases were involved. Everyone was to join up over the former Zuiderzee and head due south over the palace. Because each flight would "stack down", (fly lower than the flight ahead of them) the altitude between the first and the last element differed significantly.

315 Squadron would lead the tactical outfits with four boxes of four, so they would fly behind and below the leading Gloster-Meteors and 314, with Poop in the lead, behind and below 315.

Frits had started out as Hans de Jong's #3 man, but had traded for the #4 box position so he flew the last and lowest position of the whole parade! The sun would be just overhead or smack in his eyes on the final run. He wore his sunglasses and pulled the visor down as well. It was going to be fun.

The lead plane would set a steady course and steady speed. The second group of planes could stay in formation by making minor adjustments, the next bunch would have to make bigger corrections in order to stay in position and every change they made was murder on the flights behind them, because their corrections started to vary from "stopcock and dive brakes"" to "full power"! All Frits could hope for was that the 315 lead would figure:"The hell with the Glosters" and stay out al little further, allowing him to maintain a steady speed for the F-84's. If not, he had a murderous time ahead of him and might make a complete ass of himself!

Flying North and turning East was a breather. The flights stayed loose enough so it didn't realy call for hard work, but once they joined up with the Gloster-Meteors in a wide turn and the call came: "Tighten it up!" , the work really started. When they hit I.P., (Initial Point), two miles out from the palace, all the planes were in formation, but all Frits could see was the tailpipe and the sun, ahead and above him. He had a feeling that he was so close to the ground that he might come home with leaves in his intake, but he couldn't afford to look at anything but that tailpipe that bopped up and down above him. He sat out just far enough that Hans' tail and Frits' nose would have cleared by an inch if they had bounched into one anothers path.. By sitting off to the left a little bit he could keep Hans' tail between the sun and himself, but when the leader called: "Ten seconds to target, look good!!" he had to shift to his right about a foot in order to be directly behind his leader and was staring right into the sun. All he could do was talk to himself: ***"Keep it steady, baby! Damnit Hans, easy on me, please! Addaboy! OK, OK, steady!"***, while his left hand kept shoving and pulling the throttle constantly in order to maintain his exact position. The longest minute of his life took him over Palace Soestdijk and the A.F.B. Soesterberg as word came: "Meteors turning left!"

The parade was over and the F-84s relaxed a little bit and started a slow climb. Frits moved back to the left and his shade and took a deep breath. They survived one without scratches.

By the time they crossed the "rivers", the 1st tactical turned slightly to their left in order to return to Volkel and the 2nd tactical continued on to Eindhoven. 315 and 314 started to separate in order to prepare for their individual approaches to landing. Poop's voice sounded out: "Fancier lead to flight: all planes with fuel in tips, pull off, rest continue in formation!"

POP! Hans pulled up immediately and Frits didn't even hesitate, POP, right on his tail! Full power, up to 8,000 feet, his left hand finding the gunsight switches. He didn't even look if his tiptank lights were on or not, he had at least twenty minutes in his wing tanks, his forward and his main tank. He knew what Hans was up to and he was ready!

Hans rolled over the top and started a screaming dive, upside down, toward the little silver boxes moving slowly over the green landscape. Frits in hot pursuit, switch to "F" channel, just in time to hear Hans: "TATATATATATA" as he finished his pass and broke off to the left. "TATATATATATATA!", Frits had his sights right on the silver boxes and pulled off going after his leaders tail. Hans was back at altitude rolling over again for his next pass and Frits followed him within 20 seconds. This time their target was the 8 boxes of four, heading back to Volkel. "TATATATATATATA!", Hans again, then Frits: "TATATATATATATA", he picked off one, then : "TATATATATA" another and slid behind them diving for the ground and starting his pull-up.

Then it happened… .. the plane shuddered… … he hit the Mach… .. his maximum speed of 623 MPH… ..

The stick nearly froze, he yanked the throttle closed, popped the dive brakes and pulled on the stick with both hands and all his might! His oxygen mask slipped down to his chest,,,,,,, his eyeballs hung on his cheeks and he blacked out while his "G" suit inflated so hard, it pressed a turd out of his intestines. He expected to hit the ground in a second but hung on and as he missed the ground and started to climb and lose speed, his vision came back and at 10,000 feet he rolled over and looked down. "How close did I come to the ground? Ten feet? And how close did I come to the airplanes in the little silver boxes as I shot by them on the way up?"

His "G" meter read over 8 "G"s! Good God, the plane is only stressed for 7 "G"'s maximum! That was too close! That was also too stupid!

Hans was nowhere in sight, Thank God! "Black four,going in for landing, out!" Hans didn't reply and Frits went home.

On the ground he inspected the plane closely and found a few wrinkles in the medal on both sides of the fuselage, just befor the tail section. He suspected there were more on top where he couldn't see.

He was still shaking in his boots! "How stupid could he get?"

Watching his speed at all times was the most primary of

all flight rules and knowing that he was split seconds away from hitting the gound at full speed and that he maybe came within just feet from hitting other aircraft he couldn't see, had him shaking in his boots for hours. It was a month before he finally confided in someone as to exactly what happened . The wrinkled tailsection he had to report of course, but while the fleet of F-84-G's was being dessimated, tail sections were plentiful, Thank God!

That was the closest he'd ever come to dying and he wasn't particularly pleased with it!

Three months had passed and still no word!

Not from the Academy, not from Headquarters, not from Poublon, not from the base commander, not from anybody!

Weird! Not a word saying; "you flunked!" or "you goofed!" or "we don't want you!" nothing! As if it never happened!

As if he had never attended the Academy!

He convinced himself that it wouldn't make any difference anyway, because he had made up his mind a long time ago.

Of course, that was not entirely honest!

If he had been promoted he might have stayed in, even though he would probably say: "I stayed, cause I wanted to go through the sound barrier!"

Well, nothing happened and he submitted his final papers for discharge as of September 1, 1955.

Four more guys were leaving on the same day and arranged to meet on August 31 in Eindhoven to celebrate the sad event. Sad for the Air Force, that is.

Over the years he had never taken any vacation days, other than T.D.Y, so he was entitled to a lot of days off and he took advantage of that.

There was virtually no more flying on the old F-84's anyway, they were being transferred to Turkey and he was not chosen to help fly them over. Soesman saw to that.

Everyone else was busy with the transition to the "THUNDERSTREAK" and the "quitters" weren't invited because they would transfer to Deelen for the Piper-Cub conversion.

Hanging around home with nothing more to do than dancing, partying and hanging out at the beach, was something he could live with for the rest of his life! He made no effort to find a job, paid Mam room and board and also paid for everything he drank.

Many times he was asked to play waiter or bartender for one of the places in town and always raked in good money in tips.

The timing was great too! Mam and Pa Groot were getting married!

Pa would sell his lttle farm, pay off the mortgage on Café Rustwat and he and his two youngest daughters would move in. Mam was as excited as a spring maiden and made every effort to get Pa's kids to meet up with hers. The nice thing was, they all had a lot in common. Kees, the second oldest, played a mean accordion, just like his Dad.

All of them loved to sing and dance and one of them, Nick, even looked like Frits. They could have been brothers!

The oldest son already lived in Canada, so they never got to meet him and the oldest girl, Truus was already married. All the boys were on their own already, so only the 13 and 15 year old girls would come and live with Mam, Pa, Hedwig and Engelien. It seemed like a happy arrangement!

The wedding was great! All of Mam's brothers and sisters, all of Pa's releatives, it all added up to one great big happy family with lots of drinking, dancing and singing!

Mam said: "Frits, thank you so much, but you need not send any more money!"

He had been contributing to the family income since he was fifteen an now it seemed that he had certainly done his fair share!

It was a great life!

On August 31, just three guys showed up, Katjang, Shorty and Frits. First at the mess, where everybody toasted them and wished them good luck. Then into town. For old time sake, they had invited Beppie and her best friend Antoinette to celebrate with them and celebrate they did! Out of one bar, into the next, toasting everything in sight: the Air Force, the pilots, the girls, the bartenders, the waitresses, the musicians and the pilots again.

Obviously, they got a little schnockered! A little?

In one club, they couldn't remember what they did with Shorty and Frits tried to make out with Antoinette on the dance floor.

At one point, he thought he was making good headway

Pa and Ma's wedding

with her, but when she left suddenly in a cab, he discovered he was all alone.

Katjang had taken off with Beppie! In his drunken stupor, Frits got mad at both of them, although he probably provoked the whole situation by trying to get in Antoinette's pants and being very obvious about it.

At the Poort van Cleef, he found Shorty again and just about at closing time, Katjang reappeared, sans Beppie!

"Where's Beppie?"

"She's home, I took her home!.'

"You bastard!"

"What did you want man? Both of them all to yourself?"

"Drop dead!"

The argument continued in the car and from the back seat, Frits gave Katjang a shot in the head and within minutes, they were rolling in the street!

It didn't last long. At bar closing time the police paced the streets in their V.W. convertibles and after tearing them apart and finding out what started it, they arrested FRITS. He was a site to see, dressed in gray flannel slacks and a wild Hawaiian shirt, he stood out like a sore thumb in the back seat of the open VolksWagen.

Within the hour , Katjang and Shorty picked him up and the

police released him because Katjang would not press charges. They sang all the way to Kamp Zeelst!

It was the last time he slept on the base.

CHAPTER XXIV
DEELEN (AGAIN)

"Pull that stick back!... ... . Back!... .. Back!... .Fuck-it... Take it around!... ... ... "

Roaoaoaoaoar... ..full power... ...

Up goes the little Piper Cub, climbing at a speedy 45 miles per hour.It was Frits' third landing attempt and he hadn't hit the ground yet.

"Now goddamnit, forget your stupid jet-habits and let this thing land! On final I want 70 miles an hour, power off and this time when you round out, don't look at the damn airspeed, look outside and land the goddamn thing! I don't know where you goddamned jet-jockeys ever learned to fly or how you ever survived this long, cause you guys certainly don't know how to fly worth a shit. Now let's do it right this time. Carburetor-heat on... ..reduce power... ..right turn... ... let that speed drop to 75... ... right turn now... ..seventy... ... .at seventy... .NOT eighty... .. or eighty-five... ..SEVENTY... ... O.K... ..O.K... ... It'll fly... . Don't worry about it... ... .O.K... ... now flare out... .. O.K... .. back... . back... . O.K... ... . O.K... ... .! Now that wasn't so hard, was it? As a matter of fact, that was a good landing. Give it the gun... . carburetor-heat off... ... Let's do this twice more like that and I'll send you up by yourself!"

They'd been at it for two hours.

The First-Sergeant-Pilot had gotten madder by the minute. Frits insisted on picking up more speed than he needed. Any speed below 150 seemed like suicide to him and here he had to cruise at 75, stall at 20 miles an hour and glide at 70. It had been more than two years since he stalled an airplane and that was back in Texas in a T-28. Ever since then, his flying life dangled between 150 and 623 miles an hour. The little Piper was quite an adjustment for him as if the Tiger-Moth, the T-6 and the T-28 had never existed.

The First-Sergeant-Instructor, old enough to have been a Bird-Colonel in the U.S.Air Force, obviously hated his job and he hated all those snot-nosed jet-jockeys with their cocky attitudes that descended on Deelen as MAAND VLIEGERS. (monthly flyers)

With the departure of the old F-84G and the introduction of the swept-wing F84-F *Thunderstreak* no more reserve flying would be done on jets. Everyone descended on Deelen and it's Piper Cubs.

Life had changed drastically for Frits and anyone else who had left the service.

Frits had moved in with Leo, the bartender of the N.C.O. club at Zeelst. Leo's jolly, chubby wife Tony must have been the best cook in town and their three young boys made for a lively environment and great family to live with.

As in most Dutch homes, there was no bath or shower, so Leo and Frits built one outside of the kitchen with a coin operated gas heater. That made it much more livable for the Americanized young man and because the house was half way between the base and downtown made for an excellent place to live. That way he could still party at the club or be in town in no time. Fortunately, the four day / a / month reserve duty paid enough to cover his monthly room and board bill and all he had to worry about was his drinking money. Just flying four hours a month earned them their flight pay. Showing up for four days paid per diem, plus they received two-forty an hour for time spent in the air. Not bad!!!

Most every pilot who left the service applied to join K.L.M, Swiss-Air, LuftHansa, (just starting up), Japan Air Lines and even Iran and Saudi Arabia.

While applying, they seemed to have great jobs... ..Assistant Manager... ... Assistant Procurer... ... ... Assistant Big Shot... .. all Assistant Mr. BIG and a lousy paycheck to go along with the big title!!!

Frits wasn't about to work for anybody! Certainly not for an outfit with uniforms and dozens of superiors!

Besides that, he was pretty sure that a company like K.L.M. would take one look at his record and say: "No Thanks!" It was a good reason for not even trying.

Since he could pick and choose which days he could fly, he chose to fly early in September and all four days together.

He had always loved the town of Arnhem and he hoped that it still would prove to be a swinging town.

The first night out he first checked out the N.C.O. Club at Deelen. It was located in a German built brick building along the highway and it sure as hell proved to be like old times! Shorty, Ike, Keesje, Ari-Ari and bunches of others created a party like in the good old days!!!

Then came the first flying day!

A Master-Sergeant-Pilot, a First-Sergeant-Pilot and a Sergeant-Pilot ran the show and briefed them on the function of the Piper Squadron, the airplane and the procedures to be followed. It seemed like fun and rather lax, so they certainly could live with that.

But, then came the shock! The little Piper hopped off the ground before Frits even thought they had reached "rolling" speed, leave alone flying speed.

It flew alright. It handled alright, but the speed was so low, he constantly wanted to push the throttle through the firewall. He drove the instructor nuts and in turn, he hated the First-Sergeant by the end of the flight. The S.O.B. was so demeaning! It was as if he had to relearn everything!

Coordinated turns, stalls, take-offs and landings, everything!

The only thing that was a cinch was navigation, because they never went far from the field. They also didn't go very high!

Finally, after more than two hours he went on his own and all of a sudden, everything seemed more simple. After some additional practice stalls and a few more landings he felt more confident and relaxed with the plane.

From then on they flew mostly by themselves. Navigation mostly. It was hard to have much fun with the Piper, because he couldn't roll it, loop it or put it on it's nose or on it's tail, so he had no choice but to kinda cruise around a bit, the way the plane was designed to do.

There were a few ways to improve the dull and dreary routines of small airplane flying.

When going home to see Mama up in North Holland, he'd visit the farmers of the neighborhood by flying between the banks of the thousands of little canals that divided their properties. His wings would be about four feet above the ground on both sides of the canals and that way he was eye to eye with the enthusiastic farmers that waved at him from their crouching positions amongst the plants.

That whole area consisted of "polders", former lakes, that were pumped dry and were crisscrossed with all those perfectly straight, perfectly perpendicular narrow drainage canals.

The windmills along the dikes around the "polders", pumped the water "up" out of the drainage canals into the "ring-canal" that surrounded the old lake and in turn, that water was pumped into the North Sea.

Obviously, that left the polder way below sea level and the top of Mama's house, built along a dike, was still 19 feet below the North Sea. If a sea dike ever broke, the disaster would be a repeat of the catastrophe in Zeeland in early 1953.

After a low pass over the house, Mam, the kids and the customers would all pour out of the bar and frantically wave and shout at him. With open windows and the engine shut down, they could actually shout back and forth at one another.

Another kind of fun was: flying along a deserted river dike and touching one wheel on the dike and setting it spinning. Then hop to the other side of the dike and touch the other wheel. That way they'd fly along with both wheels spinning under them. All this at the "tremendous" speed of 80 miles per hour.

The nights in Deelen and Arnhem turned out interesting too. The first night into town, in the first nightclub he walked into, he ran into an old buddy seated with two women, right off the dance floor. Introductions went swift and things went along smoothly. Some dancing, some belly rubbing, some drinking and finally all of them in the car of the pilot buddy and "home!"

Home of one of the girls. It turned out, both of them were Army wives and the one with the home had a little baby at home and a baby sitter to pay off. They stayed until it was time to get up and make breakfast at the base at 7:30. The only drawback was: the real "doll" of the two was already taken by his buddy, so Frits had to settle for second best. Well, at least she was enthusiastic in bed. That made up for the lack of looks and personality. Another good thing about the arrangement was: they were always available on week nights! No problem!

This "maandvliegen" thing turned out to be a great innovation! It paid the bills, it provided a lot of fun and sociability and it was a break from the regular "Eindhoven" routine.

A pleasant and funny interlude was Clasien's wedding. Having all the relatives together was fun enough, but it became a wedding to remember.

Frits had been tending bar in Alkmaar the night before the wedding and was cycling home a little after midnight.

Right at the outskirts of town he encountered a hitchhiker in uniform, *his own brother Paul!*

"Paul! Hop On! How come you're so late?"

"I got restricted, so I waited till after bedcheck, climbed the fence, hitchhiked home and here I am!"

"You must have gotten lucky hitchhiking, cause you got here pretty quick. By the way, why did you get restriction?"

"Oh, that's along story, but there was no way I was gonna miss my sisters wedding, restricted or not!"

Paul had been a Marine now for two months and probably was not a real model soldier!

"Wonder where he got that?" Frits mused as he pedaled along.

The wedding started pretty normally. At 8am, ceremony at Town Hall. They were now married legally.

At 9am: Church wedding! Married in the Catholic faith in order to get God's approval.

Clasien and Ben getting married – Engelien on the right

Everything was beautiful and swell until halfway through the service the Deacon came down to Frits' row and signaled him to come out.

In the isle he whispered in his ear: "There are two M.P.s here with an order for your arrest!"

"Who? Me? OK, OK! I'll take care of it!"

In the vestibule were two big Marine MPs. Big! Six foot four! White helmets, pistols and billy clubs!

Real imposing!

"Hi gentlemen, I'm Frits Forrer and I'm sure you're looking for my brother Paul, right?"

"That's correct. He's AWOL and… "

"I know, I know! He shouldn't have done that, I know, but he didn't want to miss his sisters wedding and she and his Mam aren't even aware that he went over the fence in order to be here. So let's do this. Please, let's do this!

I'm an Air Force pilot and I'll take personal responsibility. Let him finish the church service. He's gotta go home anyway to get back in his uniform, so go to our place, tell the help I sent you, get some coffee and breakfast in the kitchen and we'll join you there in a half hour. From there I'll call your commanding officer and see if we can get some leeway, based on my personal guarantee and if not, you load him up and take him away. Fair enough?"

"Sounds fair!" A pair of firm handshakes and: "Tell them at home to feed you well, OK?" and they were gone!

To the Deacon: "They weren't looking for me, they were looking for Paul!"

"Oh, I didn't even know he was in the service! I'm sorry!"

"That's OK! You didn't know!" (everyone knew Frits was in the service!)

On the way back to the house, Frits clued in Paul and said: "Maybe I can get your commanding officer to leave you here for the day on my guarantee that you'll be back in the morning! We'll see!"

Wrong, WRONG! They were dealing with the Marines! The MP. Commandant would not assume authority, Paul's squadron commandant wouldn't hear of it and they all acted as if Paul had killed a dozen innocent women and children and was the worst criminal on earth!

Paul and the MP's had gotten acquainted and Mam and the Bride had been informed and half of the guests had to come to the kitchen and meet the two hunks in uniform! Thank God, the guys had a sense of humor and didn't seem particularly pressed to get on with things. Paul changed clothes, had breakfast with the wedding party with just one little difference: he drank beer instead of coffee or tea like the rest of the gang. Lots of beer! While Frits went from one phone conversation to the other, the MP's enjoyed the hospitality in the kitchen and Paul played piano with everybody singing along! Sang and drank and drank and sang! By eleven thirty, the MP's didn't have a choice any more and moved Paul out under loud applause and laughter. Outside they posed patiently for pictures and it became indeed hilarious, Paul, drunk as a skunk, beret askew on his head, held up by two strapping, smiling men in uniform. The whole wedding party waved them goodbye, while singing: "He's a jolly good fellow!!!!!!"

It was the highlight of the day!

Meanwhile in Eindhoven, Frits had to face the cruel world and decide on some sort of income.

He'd gone home and worked for Mam in the bar and

Paul between two MP's

worked as a bartender or waiter for other friends in the bar business, and always made good money at it. After all, he grew up in that business and like his brothers and sisters, was fast and efficient. They knew how to move beer, drinks and money at a fast clip.

Meanwhile he looked at a few other things to do but always with the distinction: he'd have to be his own boss.

One evening in the Cinzano bar, Beppie introduced him to a gent named Willem from Roermond who dealt in Frankfurters. Frankfurters in jars, 10 to a jar. He'd like to develop a Eindhoven distributorship and Frits might be his man.

O.K.! O.K.! A trip to Roermond, a tour of the plant and a visit with the "Old Man", Willem's father and the owner of the plant.

A business plan was worked out. Frits would be the distributor for a fine tasting frank, produced by a fine, small, unknown but solid firm.

Transportation became a must! Scanning the papers, Frits found an ad for a '48 Olds in good condition and reasonable price because of it's gas guzzling ability.

A loan arrangement was worked out and pretty soon Frits had an easier time than anybody finding his car on the crowded market square. *It was taller than any car out there!*

Building a clientele was a matter of hard work. Calling on bars, corner food stands and restaurants and sampling the "bigger, better" frank took time, but it worked.

He had stiff competition. A local soccer hero was already well established with a competitive line and he commanded a lot of loyalty. Little by little, the business kept growing, though.

On one of his "maandvliegen" trips he left early and dropped by the Sergeant-Major in charge of food purchases at Deelen and presented his product.

The Sergeant- Major didn't even want to taste one. "I'll need two per person for the Officers and N.C.O. lunch menu, ten days from today and one each for the enlisted men the next day. Let me type you a purchase order, I'll be but a minute. If I hear one complaint, I won't order any more, O.K.?"

"Can you believe this?" Frits was on the phone with Roermond, the big boss himself.

He presented the good new: *"More than a thousand jars!"* and the problems: *"How do I get it here and we'll have to bill the government and where do I get the money to pay for it in the mean time?"*

"No problem!", The big boss sounds reassuring. "We'll drop ship directly to the base and WE'LL bill the government. When their check comes in, we'll credit your account with the profit. Don't spend it yet, but boy, you made yourself four hundred guilders today! Get some more accounts like that!"

"WHOOPEE!!!! Now that was one hell of a business.!"

While celebrating at the club, Frans, the bartender, told him that the military clubs bought their own supplies and he told him which sergeants and officers he had to see in order to get his franks into all the clubs as well! This could turn out to be fine deal!

Back in Eindhoven he had to deal with a major who wanted samples dropped off in the kitchen for testing by the chef. If the chef decided that the higher quality would justify the higher expense, he would put them on the menu on a bi-weekly basis. The chef approved and Frits moved to other military camps and bases, but not always with the same positive results. At Volkel, for example, he could never get to talk to the right person.

The most positive thing about the Air Base business was:

it gave him a basic monthly income that allowed him the time to go around seeking new customers.

The most negative aspect was that he didn't have to work too hard any more, so if he made a round of deliveries and by 10 o'clock in the morning he encountered some acquaintance in a customer's pub, who challenged him to a game of billiards, it would spell the end of his working day. He'd roll up his sleeves, order a beer and started one game, got challenged into a "revenge" and then another and before long he'd have too many beers to call on customers, so he'd start chasing girls instead of customers.

Many a day he drank up more than he took in, but man-o-man, was he having fun! Not a care in the world.

The crew room on base became his customer, which was a great reason to drop in and shoot the shit and reminisce with his old mates. One of them, Katjang, had not resigned, but signed up for another year. That Quisling ! That traitor! That Benedict Arnold! Well... .they stayed buddies just the same!

As his route kept growing, he kept getting requests for other products. The Dutch love their "croquettes" and "meatballs", so Beppie introduced him to a skinny butcher in town by the name of "Adje", who had a fantastic "croquette" mix. He also had a reputation of buying cheap meat from neighboring farmers, which wasn't quite legal, but very profitable. They concocted a business venture of supply and demand. Adje would supply if Frits would demand within a certain time frame. That aspect of the business was not a wild success, but Adje was such a delightful nut, that they laughed their heads off, most of the time.

At one point, Adje and his wife, Frits and Beppie were going on a trip down the Rhine in Germany. Frits was driving Adje's car and when they got to the border checkpoint, Adje discovered that his passport was still on the dresser in his bedroom. What a bummer! They had driven an hour and a half just to get to the border and now they faced a three hour trip just retrieve a lousy passport! What a bummer!

NO WAY! Frits had grown up along the border further north and he knew that many farms straddled the border and that dirt roads sometimes went unguarded. So, they turned off the main road and tried to follow the border south on secondary and even dirt roads. Sometimes they weren't quite sure whether they were still in Holland or not until Frits made a sharp left on a muddy trail. Bouncing across a fairly firm cow pasture and through an open gate at the far end landed them on another dirt road. He was *pretty* sure that they were in Germany by then, but they were *very* sure when they passed a pair of German "Zoll-Beamten" (border guards) on their bikes.

When the border guards realized the Dutch license plate, they whistled and shouted: "Halt! Halt!"

"Like hell! Up the "halt" mit you!"

Instead of halting, Frits stepped on the gas and drove as fast as the dirt road allowed him with the women screaming in the back seat and Adje applauding and shouting encouragement in the front seat. He made constant turns... east... south... .east again... .south again... , realizing that the guards would call ahead the moment they would locate a phone and get other guards to be on the lookout for them! When they came upon a highway, he turned south until they intersected an east-west road and followed that east until they hit the autobahn to Cologne. They finally sighed a sigh of relief and the women finally calmed down some. In the first pub they encountered in Cologne, they drank to their "invasion victory" and also made sleeping arrangement with the friendly owner who rented sleeping spaces upstairs. Instead of following the Rhine further down, they decided to stay in Cologne and explore the Ancient City. And explore they did! They partied in a dozen Café 's, nightclubs and restaurants for two days until it was time to head back to Holland.

At the border, no one looked for passports and all they were asked was :"Anything to declare?"

"Nothing!", of course!

Beppie's sister Truus had bought a little bar across town and because Beppie and her stepfather Jan kept getting into hassles all the time, Beppie moved in with Truus, but kept working in the Cinzano bar with her mother. Frits mostly slept in his own bed with Leo, Tony and the kids, but sometimes he slept over with Beppie and that way he maintained his freedom.

In November, he celebrated his birthday at home with the family in St. Pancras, and even though it was a great party, it didn't even resemble the brawl of the last year. But... ... .. Life was great!

CHAPTER XXV
THE FUN OF FLYING!

Though the little Piper Cub could hardly be compared with the sophisticated Jets of that era, it was still a machine that managed to get off the ground and stay there for as long as the pilot and the gas supply decided.

In other words, it still provided some form of flying and a form of fun.

Some of the fun was provided by the Army.

The pilots of Deelen were assigned to fly "forward gunnery controllers" over gunnery ranges where the controllers directed the fire of the Army artillery. By watching the impact of the hits, the airborne could radio back to his unit how to correct their aim, depending on whether the hit was in front, behind, to the right or to the left of the target.

For the pilots the fun was the freedom they had in flying the plane. Their passengers had no knowledge of flying at all and believed everything the pilot told them or did. That allowed the pilot to fly between the trees, (to avoid detection, of course), buzz the artillery, (to check on the troops, of course) and hop all over the country side if he felt like it!

The attitude of the Army Officer also effected the flight.

One Officer might be a Colonel who'd shake hands and say: "Hi Sergeant, I'm Joe Blow and I haven't done much flying in my time, so can you kind of take it easy on me? I understand you guys can make it as rough or as smooth as you want to, so please, smooth for me! O.K.?"

That attitude would earn him a smooth and exciting ride.

Frits would treat him like a student and explain every detail of the walk-around pre-flight inspection and the pre-start-up procedures. As they taxied out, he'd have his passenger feel through on the controls with him and on takeoff, he had him pull on the stick with him and once off the ground, he'd announce: "You're flying!"

He'd guide him through easy turns, climbs and glides and usually had a very enthusiastic passenger on his hand who envisioned telling his wife and kids at night: "I actually flew the plane!"

Depending on the fortitude of the passenger, the flight would be very smooth and easy or wild with steep dives, steep turns and low passes.

One way or the other, the passenger would thoroughly enjoy himself and thank Frits profusely.

Another type of passenger was the cocky First Lieutenant who'd snap a sharp salute: "I'm Lieutenant Doe. What's your name and where's your cap?"

"I don't fly with a cap and my name is Frits Forrer!"

Frits would look "fairly" unmilitary with a loose hanging flight suit without insignias or rank indications, wearing "very" unmilitary loafers and his purple and white striped scarf draped around his neck.

"What's your first name?"

"Just address me as Lieutenant!"

That kind of guy would be in for a rough ride!

Sudden steep turns and climbs would have the lieutenant grab hold of his seat for fear of falling out.

Getting him to throw up was a cinch!

All the pilot had to do was push the stick forward suddenly and the rear seater's stomach would land in his throat.

Just a few: "Lieutenant, do you see that?" A hard push forward and a slow pull back.

" Lieutenant? See that there?" Push hard, pull slow.!

"See that, Sir?" Push and back up! "See that, Sir?" The lieutenant would gag and throw his guts out! The poor bastard in the back seat would be as sick as a dog, but the mission had to be completed, so as sick as he was, he would still have to pinpoint artillery impact and report his corrections back to base. All this time he'd have to sit in his own stink while the pilot in front simply let the fresh air blow in through an open window.

Upon return to the air base, the rules of the Air Force applied: "You carry your own parachute and you clean up your own shit!" So the passenger cleaned up his own mess or paid the mechanic five bucks for doing so. That type of passenger tended to lose some of their cockiness around pilots.

Other fun flying was delivering supplies or a passenger to a small field where troops were on maneuvers. Sometimes the field was the size of a football field, so they had to drop in over the trees at near stalling speed, just hanging by the prop. When it was time to get out, the ground troops would pull the fuselage all the way back between the trees in order to increase the take-off run. While the troops held the tail off the ground, a few guys would grab the struts and as the pilot gave it full throttle, the strut walkers and the tail holders would push and run as fast as they could to give the plane a good head start. It was not quite like being shot off a carrier, but it helped. Keeping the plane close to the ground to gather as much speed as possible, the pilot would pull the plane over the onrushing trees at the last second, clearing the tops by inches. If they were flying the "Super-Cub", they'd pull full flaps at the last second and the plane would rise like a helicopter. Every time he'd clear the trees, he'd sigh a sigh of relief! Survived another one!

Part of the "reserve deal" was a two-week training exercise, which meant: sort of a vacation, away from work and wife for most of the guys. Most of them had regular jobs and just came in to fly on Saturdays, just one day a week. They were not involved in the partying and the camaraderie of those with less stringent ties or no ties at all.

As a result, "two weeks on maneuvers" had a party and vacation atmosphere from the start.

The first one Frits was involved in brought together about 40 pilots, headed up by a reserve First Lieutenant, who was a young lawyer in civilian life. He really thought of the whole affair as a "vacation trip". The show was actually run by the Sergeant Major and the Sergeant, who ran the briefing and explained the mission.

The whole group was to fly in formation,,,, (What? Formation?") to a civilian airstrip in the west end of Noord Brabant and they were to be attached to the Special Forces Unit in Roozendaal for support and gunnery control.

"*WOW! ROOZENDAAL*! Great town!"

They were to camp on the field in tents and fly an intense and varied schedule of activities.

The Sergeant-Major would lead the flight and he gave the breakdown of the flight-leaders, element leaders, wingmen and passengers.

A wide left hand turn should get all the planes in formation and on their way.

The weather was fairly shitty. Clouds at 1200 feet. Visibility; five miles. Windy, gusty and bumpy ! "SO DO THE BEST YOU CAN!"

Well, that gave everybody a lot of leeway.

WHAT A SCREWED UP SITUATION!

It was so bumpy, it was nearly impossible to fly formation with *one* plane, leave alone with *four* and it was certainly impossible to have one formation of four, fly formation with another formation of four!

The wind popped the little planes all over the sky and the main concern was NOT to fly formation, but to avoid mid-air collisions at all costs. The important thing became: Fly in the same general direction! Miracle of miracles! They all found the field in a drenching downpour without any collisions and all of them landed without incident.

Tents had already been set up, chow was ready for eating and because of the rain, all they could do was sit through one briefing after the other until it was time to eat again.

No transportation was available for trips into town, so it was expected that the pilot would camp out and rough it.

Wrong!

Like any decent airstrip in the world, this one had a pub across the street from the field.

Slightly soaked, but very thirsty, they arrived at the pub and started to lift their steins and their spirits. In no time flat, the joint became a madhouse of songs, jokes and laughter.

The "base" was guarded by a unit of the "special forces",

Piper Squadron on maneuver

identified by their purple berets.

The man in charge, the "wachtcommandant" , Sergeant Wil Bonder was a career soldier who could fall out of the back of a speeding truck as easily as he could down a dozen beers.

He had never been associated with pilots before, but he liked what he saw!

The little barroom was crowded with singing and laughing young men, who didn't seem to have a care in the world. It didn't take long before a friendship had sprung up between Wil and Frits and not much later they were off "nightclubbing" in Roosendaal. Girls were aplenty and even though a sloppy flight suit and heavy boots were not exactly the glamorous attire for dancing, Frits danced his butt off. Somewhere during the night he lost it! Too many beers probably. The last he remembered was snuggling with a redhead on a couch and a beer in his hand.

The next thing he remembered was waking up in a typical military bedroom, still dressed in his flight suit and a mouth so dry, it felt like he had swallowed the entire Sahara desert. His watch said: 8:30 and it was broad daylight. He spruced himself up as best he could at the corner sink and found himself in the officers mess filling up his tray at the buffet. There were just a few officers still eating and they, nor the kitchen help asked any questions from this strange looking individual in a flight suit without markings or rank, high boots, a purple scarf around his neck and a purple beret sticking out of his breast pocket.

He ate and drank heartily and then meandered across the compound of this large "garrison" until he located the motor pool. An old Warrant Officer didn't even question him when he asked for transportation to the air strip and simply complied with his request.

The only one who questioned his 9:30 arrival was the Instructor-Sergeant and he shrugged him off with:" I was doing liaison work with the special forces."

"But Forrer, if we hadn't been so socked in by the weather, you would have missed the morning maneuvers and.."

"Sarge, you'll just have to learn to win this war all by yourself!"

With that, he headed for some coffee.

The weather remained stinko, but at least the rain had stopped. That opened some new prospects.

The little airport doubled as a racetrack and almost the entire perimeter consisted of a huge oval, marked off by old tires. Thousands of tires, stacked two and three high around the entire inside and outside of the muddy track. Apparently they doubled as markers as well as crash walls. Anyway, the tires spawned a new idea.

With a bunch of tires a 40 feet circle was created near the tents and the pilots were divided into "pilots" and "bombardiers."

With the clouds hanging at about a 100 feet, it was decided to make three bomb runs from south to north at 50 feet. With the window up and the door down, the "bombardier" could hold a tire outside the plane and he would drop the tire while flying over the circle. The team with the most "hits" would win a case of beer and the entry fee was just one buck per person.

Two tires were stored behind the "bombardier" and with one in hand, they took off with one minute intervals. The pattern was quite simple. Climb to 50 feet, turn left, fly past the airfield, turn left, and another left, lining up with the target. The entire ground crew stood around the circle, scoring and cheering. One by one the little planes came over and proved how stupid and inefficient they were. Everything went beautifully except for one. The stupid tires bounced three or four times before stopping and being released at 70 MPH, they ended up 100 yards down the field even if they initially hit inside the circle.

So the trick was: drop them early and hope they would end up in the circle somehow or make them fall flat and hope they wouldn't bounce that way.

Only one ended up in the circle on the first run and that was by accident, the bombardier dropped it too soon!

Great technique was applied on the second run. Some planes came over at 40 mph with full flaps, hoping to improve their accuracy. The only thing it accomplished though was stagnation in the pattern as the planes started to catch up to one another. It didn't improve the "hits" at all. On the first run, the planes came over in an orderly fashion with one minute intervals. On the second run planes were closing in on each other and started to scramble out of the pattern with the result that the third run was complete chaos. Planes all over the sky! Some of them just landed, rather than chance a collision, other tried to approach the circle from a different direction when the coast was clear.

It resembled an air battle of World War One.

The bombardiers tried early release, tried to spin them in a flat spin, so they would land on their side, but when it was all over, only three tires ended up in the circle and the three winners had to split the case of beer. HOORAH! Another excuse to celebrate!

Weather cleared some eventually and the Army assistance maneuvers started up.

Frits got permission to take Wil up for an "inspection" ride which included buzzing the apartments where Wil's girlfriend lived and buzzing a tank formation , camped out in the woods.

Through some of the clearances, they could dive down right on top of the tanks, barely missing their long and nearly invisible antennas. The Army guys threw cans and bottles up at them, but when the guys on the ground learned to "lead" the plane, it was time to break off the attacks. They barely missed a coke bottle that someone threw up well in front of them and that was too close.

When the "Purple Berets" threw a party at the Garrison for the Air Force, but cut off the beer flow at 11 o'clock, the disgruntled Air Force guys climbed back into the trucks the Army had provided, singing: "We gaan nog niet naar huis... .!!!! "We're not going home... .!!"

The following day, six Pipers in formation dropped toilet rolls over the Kazerne and the paper fluttered down beautifully over the entire Army camp.

Some shitty message from higher up cancelled the proposed formation flight home. Instead it would be a navigation exercise, with the little planes taking off 3 minutes apart. The passenger on the way out would now be the pilot and vice-versa.

By 3 o'clock on Friday afternoon of the second week it was all over and the guys scattered all over the country on their way home. Nobody even hung around to party anymore!

CHAPTER XXVI
UPHEAVAL

A lot of things happened in fast order.

The Olds broke down. The car had been great even during the coldest winter days. It always started, it always ran and then one morning… . tok, tok, tok, tok,… . it blew a rod bearing. Frits was halfway between Eindhoven and Deelen when he blew the engine. That was the end of it and he hadn't made the final payment yet.

Things still muddled along, business wise. He made his local deliveries by bike and the bases he could handle by train and bus.

Then came a real bombshell: *Les Halles* in Paris, the great depot, went on strike and all deliveries to France stopped dead. Holland's meat surpluses soared, prices dropped and the big sausage producing companies dumped their products on the local markets at highly reduced rates. Everywhere he went it was the same story!

"Frits, we're sorry, but we got "UNOX" sausages for half price, but stay in touch, things may change!"

Or: *"Zwanenburg made us such a good offer, but check back again!"*

Other than a few loyal customers, he lost his entire business over night. That was too much!

He'd always wondered if he should go back to the United States and that did it!

He filed the necessary paperwork and was told: "There is a ten year waiting list, so be patient!"

What a crummy deal!

Other things developed, though.

On Easter Sunday he dressed up in his best suit and decided to go to a "Thé Dansant" at 4pm.

Maybe he'd meet a new girl, dance a little and make it back for supper. Tony was preparing a whale of an Easter meal and he promised he'd be back in time.

Being a little early, he decided to have a drink in the "Tonjo-Bar" next door to the dance club.

The Tonjo-bar was an old hangout and before long he was playing cards at the bar for a round of beer. While playing, he had the feeling that some eyes were piercing his back. Sure as hell, Jerry Deerenberg and Andre van Zon were sitting at a corner table with a pair of blondes.

"HI GUYS!"

He finished his hand and wandered over and got introduced to the gals.

Both the guys were "Maandvliegers" also and both the girls were locals. Both were blond, blue eyed and pretty, but in different ways. One had very "lady-like" features and manners, but the other one had a million curls, a devilish twinkle in her eye and the cutest little dimple in her chin.

Interesting! Very interesting!

After some small talk, he returned to his card game and when he got up to go dancing, they were gone.

Later during the week he received a call from Adje, who was stranded in Arnhem with his wife, because their tour bus broke down and he would rather not attempt to make it home by train. Would Frits kindly get his car and come to Arnhem and they'd stop for dinner somewhere.

Frits had to make just one stop in town and parked Adje's car on the market square. He completed his delivery and nearly got run over by "dimple-chin" on her bike.

"Hiya Frits! How are you?"

She seemed very pleased to see him and he wasn't any less

excited! He had to cut the niceties short even though he felt like buying her a beer and getting better acquainted.

Instead he said: "Hey, I've got to run to Arnhem and back. Wanna come? We'll have dinner on the road somewhere!"

"Oh, sure!"

Dimples parked and locked her bike and hopped in for the ride.

She was as delightful as she looked. She chatted endlessly and he learned a lot about her on the trip. Apparently, their attraction had been mutual. The more the guys had put him down as a "womanizer, party goer, hot shot, playboy", the more intrigued she became and the more she wanted to meet him. That was funny!

He also learned that her sister was engaged to an F-84 pilot at Volkel who got killed and through her sisters relationship she had gotten to know a lot of pilots. It sure was a small world. They shared so many acquaintances, that it was uncommon.

They were in Arnhem before they knew it and with Adje and his wife, they dined and danced the evening away. What a sport! Thank God, neither Adje nor his wife asked any questions about Beppie, so the evening was perfect.

Back in Eindhoven, dimples was returned to her place. She rented a room from a pair of old ladies and Frits could kiss her goodbye in the hall and that was it!

Yet he was on cloud nine! She was the greatest girl he ever dated and they had already planned to meet the next night.

Through dimples he met a string of new people in Eindhoven. "Phien", a tall striking dark haired beauty, married to a maandvlieger from Volkel and "Karel and Riet van Deelen", a couple with their own home. "*OWN HOME*" was significant because after the bars closed there was a place to party and... ... .sleep! Karel had to get up at 6:30 and so did some of the others, but that never kept them from picking up a case of beer on the way home and continue the party at the house. The group started to include Katjang and Shorty whenever he was in town and some other pilots as well and the parties started to become routine.

Frits and Dimples saw one another almost every evening and sometimes also for lunch, but their was no sex between them. Dimples was 21, well built and passionate, but she drew a line!

Dimples, Frits, Shorty, Appelo and Vendrig

Strange but true, Frits hung in there anyway!

He changed his flying routine to Wednesdays only . Dimples would see him to the train and the bus would take him from Arnhem to the N.C.O. club at Deelen by 10:30 at night.

Just a few nightcaps with the boys at the bar and he'd hit the sack, fly the next day and Dimples would meet him again at the 7:30 train.

He was so straight, he couldn't believe it himself! Neither could anybody else!

He took her home to St. Pancras and the family just loved her. Mama had some objections about her not being Catholic, but thought she was great otherwise. He had turned a new leaf.

He spent all his spare time with her and when sex finally came, it came very suddenly and very tenderly.

She had planned it and he accommodated her. Gladly!

He had never been with a virgin before. A virgin at nearly 22? He could hardly believe it!

The best thing about it was: *She loved it!* No, she craved it! Once she had the taste of it she wanted to do it morning, noon and night! *He* didn't mind at all.

His sister Hedwig, now 18, came to visit them one weekend and was promptly offered several jobs as barmaid and she accepted one of them. The job was as a barmaid at the "Black

Ankie and Frits at Joop's wedding

and White" bar on the outskirts of town and included room and board. It was a nice neighborhood pub and the owners were like a Mam and Pap to her. She now became part of the regular party crowd and it promptly created one drawback. Very unexpectedly and very undesirably, as far as Frits was concerned, Hedwig went *Ga-Ga* over Katjang and he was about the last playboy he wanted his sister to associate with. He did his damnedest to keep those two apart, but he didn't feel too successful.

Dimples took him to meet her mother one Sunday. They arrived by bus and enjoyed an excellent dinner in the neat little house of her widowed Mam. After dinner, the three of them relaxed and Mam felt that was an opportune time to express her opinion. *"Dimples is NOT to marry outside of her faith and if she ever marries you, I will kill myself!"*

Frits was speechless!

Dimples wasn't!

"Come on Frits, we're leaving this instant and Ma, I'm going to marry who I want to and if you decide to kill yourself, that's between you and God! Let's go!"

OOPS! That was not such a great success! Dimples was determined though and they continued their affair as heatedly as before.

When they attended the wedding of his stepbrother Jan, Frits announced that he and Dimples would be the next couple to marry in this family.

It didn't work out that way, though!

Several things intervened.

First, his Brother Joop was going to marry that July.

Second, Frits' emigration papers came through!

Already?

Since he was born in Belgium and Belgium had NO waiting list, American immigrations laws put him under a Belgian quota and his papers were O.K.'d as of September 10.

"WOW!"

Dimples and Frits did some serious talking about going to America and starting a new life for themselves over there.

Obviously, a lot of things would have to be resolved. A Visa for her, marriage before they left, probably, sponsorship for her, etc. It could all be done. He was sure of it.

Then a bombshell!!!!

One Wednesday night, after flying, his train from Arnhem

was delayed because of fire on the tracks and he missed his connection for Eindhoven in 'sHertogenbosch.

He called Dimples while waiting for the next train, but she had already left for the station. He wasn't particularly worried, because he was sure she would hear the news of the delay and be there at the next arrival.

WRONG!

No one in sight! He checked the regular pubs... nothing!

An hour later he went to the station again, figuring she might meet the next train! Nothing.

He called her home again!... Nothing... . Nada... Niente... Zilch...

Back to the Poort Van Cleef, where she most likely would look for him.

A call to Karel and Riet... . Nothing! Hadn't seen or heard from her!

Finally, after midnight he went to sleep at Karel's place, because Dimples was liable to drop by there.

He spent a miserable night. At lunch he waited for her at work and her first words were: "Where were you last night?"

"Where was **I**? Where were **you?"**

"I was at the station while you were out gallivanting with who knows who?"

"Gallivanting? The train was delayed because of a fire and I missed the 7:05 in Den Bosch, so I arrived an hour late. I looked all over town for you all night and I stopped calling when I woke your landlady at 12:30. So where were you?"

"Well, Phien said you probably... "

"You came to pick me up with Phien?"

"Yes, and "

"And she and you went gallivanting because you thought I wouldn't show?"

"She said... .."

"She said what?"

Dimples burst out crying. "I'm sorry, but she said you were probably out in Arnhem with one of your old girlfriends like you always used to and... a and... and.."

"And you believed her? That's the kinda trust you have in me? That bitch! She's married and screws around on Bill and then she accuses me of the same thing? That God forsaken bitch! I can't believe you believed her!"

"I'm so sorry Frits, believe me, all I wanted to do was run back for the next train and call Karel or the Poort to see if you had been there , but she said: "You're wasting your time! Oh, God, but I missed you so... ." More tears... ..

They made passionate love along the canal instead of eating lunch and all seemed to be O.K., but something had snapped!

He had planned to be true to her forever and she knew that! She had to know that! She had to have felt that!

Yet at the first little delay, she distrusted him!

He had never felt so shattered in his life.

That afternoon he made a decision. He was going to America as soon as feasible after September 10th and he'd go alone.

He'd start life all over on his own. He called Rotterdam and tentatively booked passage for September 27.

He told Dimples nothing!

He called Joop to tell him to arrange a date for him at his wedding and changed his flight schedule. He'd fly on Saturday and continue on home from Deelen and he'd be Joop's best man on Monday.

Dimples felt a coolness set in. They still saw one another nearly every day and they still made love a lot, but hiatuses started to appear in their schedule. He wouldn't show up for lunch. Dimples had to call Tony and Karel, the Poort and Hedwig trying to find out where he was.

He didn't talk America and wedding anymore and when she mentioned Joop's upcoming marriage he simply said that plans had changed and he did not elaborate. She was still fun to be with, but when they started to drift apart, little by little, she started to spend more time with Phien and that certainly clinched it for Frits.

If that's who she wanted to be like, more power to her!

CHAPTER XXVII
WINDING DOWN

Joop's wedding nearly did get screwed up!

Not actually Joop's wedding! No, just Frits's presence at the wedding!

He went out to party in Arnhem on Friday night and had to fly at about 10 on Saturday morning.

Just a two-hour cross-country and then he could probably leave for home.

They gave him some shitty assignment, like: 100 miles north at 3500 feet, then south-west to "blankety-blank" and then back to base. Horse shit!

The minute he was out of view, he turned east toward Germany for some low flying. The Germans apparently never reported anything, so that seemed pretty safe. He futzed around over Germany for about an hour and started to work his way back home. Streaking low over a large wheat field, he could cut a path through the tall wheat with his propeller and two traces with his wheels. It looked like a huge wheat field with just one small cut through it with some trees and shrubs. Too late did he realize that the trees and bushes were along a dirt road that cut through the field. The dirt road was no problem, but the phone lines that suddenly appeared before him certainly were. The poles were hidden by the trees and all he could do was pull up as hard as he could while shoving the power full forward!

He caught all four wires going straight up!

Three of them snapped immediately, but the forth one caught him like the arresting cable on a carrier! The plane damn near stalled out and he pushed the stick forward with all his might when that line snapped too!

His heart was in his throat and his pulse raced at a hundred miles a minute. "Height! I need Height!" A steep turning climb got him away from the area in a hurry. He wasn't sure whether he was still over Germany or over Holland already, but he made sure that he followed the border north for a while, so he could approach the base from the direction he was supposed to have flown to according to his flight plan. Right now he was still south-east of the base and that was not something he wanted the base to be aware of. The next thing was: "damage assessment". The plane flew alright, but he seemed to hear funny noises so once he was over 3000 feet, he pulled some stalls, simulating landing approaches and everything was near normal. The plane wanted to drop off it's left wing some, but some planes just do.

A little relieved he continued his way home, fully expecting a pair of M.P.s with their flashing lights on their jeeps to be waiting for him at the end of the landing strip.

Mentally he said goodbye to the Monday wedding, forget America in September and "hello" jail term.

He didn't feel particularly jolly and kept tossing around in his mind what he should do, once he hit the ground.

He could have saved himself the worries! No M.P.s!

He landed a half hour late and being that it was lunch time, things were really quiet on the ramp.

As usual, a mechanic ran out to put down the wheel chocks, the moment he parked the plane. The man didn't do ar say anything unusual and started to walk away when Frits whistled him back. The damage was very minimal, but very obvious!

One wire had caught right under the prop and the cowling had a perfectly straight horizontal dent in it. The wheel struts were scraped bare along the leading edge and glistened in the sun. The little wing like structures on the struts were dented, but that was all. The prop had nothing! Nothing! The prop must have windmilled right over the wires! What a relief!

The mechanic told him: "Just go to lunch. Don't check in. That gives me time to bang the dents out and slap some paint on them and don't worry! I'll take care of it!"

At lunch there was a call for Mr. Forrer! His heart stopped!

"Why didn't you log in when you landed? We've got people waiting for the plane!"

"I'll be right over!" Thank God, it was just an annoyed Sergeant-Major and not some M.P. official.!

To Frits' eye, the camouflage touch-up was obvious, but nobody else said a word.

The paint was still wet, but somebody took the plane up and Frits could breath a lot easier.

A couple of packs of American cigarettes took care of the mechanic and when he asked the Sergeant-Major: "Any chance for another flight?" he grumbled; "Get the hell home!"

"Have a good week-end, Sir!"

Joop's wedding was a charm.

Frits' date for the day was Joop's 19 year old sister-in-law and all he could say was: "Where the hell have you been all my life?"

Forget Dimples! This was beautiful and charming and it signaled a good omen! Everything about the wedding and the ensuing party went beautifully! Life started to get brighter again!

In Eindhoven, he started to wind things down. He got rid of his inventory and checked out of Leo's and Tony's house, but was always welcome to come and have supper with them or stay over.

One sunny Saturday when he dropped by, Tony insisted on him sitting down and eat something even though she was just clearing the table. They were happily chatting along while Frits was ladling tomato soup when the sound of a jet fighter ended with a tremendous explosion, that made the windows rattle and the soup jump out of his plate.

"A PLANE! A CRASH!"

Frits scrambled to call the fire department and flew out of the door with Leo on his heels.

A right turn at the end of the block and there it was, half way down the street!

One house was demolished, maybe two, but there were no flames. Out of all things, the police and fire department were already there and distant sirens indicated there were more on the way!

Hedwig and Frits

Joop and bride, and Ankie and Frits

Frits identified himself as a pilot and volunteered to help while Leo told the cops, he worked on the base and probably knew the pilot, but they were politely and firmly rebuffed.

"Our first concern is that it doesn't explode, so stay back!"

"It didn't explode?"

"No, but nobody could have survived this! Help me hold the people back!"

One thing was obvious, from the wreckage he could tell it was a T-33, but there was no way of knowing if there were one or two pilots in the wreck and how many people were in the remainders of the house?

In minutes, the police and the fire department had blocked off the street and there was little left to do, but meander down the street to go see Ari-Ari, who had just gotten married and moved in with his in laws. Ari-Ari was one of several pilots who didn't like civilian life or couldn't hack it and had signed up again and were being retrained on Jets at Volkel. As they approached the house, a girl came flying out and into Frits' arm, sobbing: "Frits, it's Ari, it's Ari!" She was Ari's sister-in-law, lovingly known as "ZWANENDONS." (Swandown) She cried hysterically while Frits kept yelling at her: "No, No, it isn't Ari! He's too good a pilot! He'll never crash!"

"Frits, you don't understand! He called up from Volkel that he would solo a T-33 again today and he wanted us to watch for him!"

"This close to the Eindhoven runway? You're crazy!"

"No, Frits, it's true, it's true, it's Ari!"

"Good God!, I can't believe this! Not Ari!"

"Yes and she is pregnant, you know?"

"No, I didn't know that either! Where is she?"

"The doctor from the base gave her a shot and she's asleep."

"Good God, not Ari-Ari!"

All Frits could think of was Ari strumming his guitar and singing: *"Don't The Moon Look Lonesome Shining Through The Trees?"*

"You going to the funeral?"

"No more funerals for me! From now on, only weddings and baptisms. Only positives! How many damn friends can you bury before you're thirty? Give me another beer!"

They were sitting at the bar in "Le Chat Noir". (The Black

Cat). Dikke Willem had built a cozy little pub in the back of the Poort Van Cleef and Katjang was the bartender.

Katjang had signed up for one more year and got checked out on the F-84F and according to rumors, he was found to be "not fit" for squadron life. He was a 2nd Lieutenant by now, but he was told that his lusty life style was a bad influence on the younger pilots and therefore he was made a test pilot at Volkel. He worked all alone with the maintenance crew. When a plane was repaired, it had to be test flown before being returned to it's squadron. He seemed to love it, but it was unclear whether he chose to leave after his year was up or whether the Air Force asked him to. One thing that the Air Force did not appreciate was a beautifully performed roll right after takeoff!

Anyway, he was now the center attraction at the "Chat Noir" and for the pilot gang, it had become the logical place to congregate.

The exact reason why Ari-Ari had managed to dive down into those houses had never become clear and luckily only one old man was killed on the ground.

It could have been worse! A lot worse! He could have wiped out entire families! If the plane had burned, he could have burned down an entire block!

"They virtually lived on the perimeter of the Eindhoven Air Base. The tower could probably read his number. How did he expect to get away with that?"

"When people are in love, their brain goes out of gear. They lose all common sense!"

"Tell me about it!"

"Frits, what do you know of love? You think that love and sex is the same thing! It isn't, pal."

"It isn't? You better give me another beer, cause this is heavy! This requires some serious drinking!"

The trips to Eindhoven were mostly party oriented from then on and he spent most of his time in Noord Holland, tending bar from time to time and spending most of his time on the beach during the day and in dancings at night.

He had gotten into "Cuba Libres" (rum and coke) and he carried a pitcher full to the beach for easy sipping. Of course, Mama didn't approve of his lifestyle, but he shrugged it off.

"No point in looking for a steady job with just a few months to go."

When one of his neighbors suggested a trip through Europe in his Volkswagen with his skinny friend, his only question was: "When?"

Mama whispered: "These guys are funny! You don't wanna go with them!"

He went anyway.

It was pleasant enough through France and Monaco, but in their tent in the hills above Geneva, the two friends did become aggressive and they parted company. The V.W. continued through Switzerland and Frits hitchhiked south over the Brenner Pass to Italy. He was down to 25 guilders, but that didn't bother him too much. He slept by roadsides, haystacks, orchards and occasionally a bed.

He toured Milan, Venice, Padua, Florence, Sienna, Rome and Naples, hiking his way from town to town. In Naples he caught the ferry to the Isle of Capri. And after spending the equivalent of one American quarter on a monastery bed, he decided to cut out that kind of luxury. He slept on the beach at the Marina Piccolo on the far side of the island and in fishing boats at the Marina Grande on the side facing Naples.

When he was down to just "ferry money", he took the ferry to Sorrento and there he was; a thousand miles from home without a penny. So far he had subsisted on bread, apples and grapes, but from here on in it would probably only be apples and grapes or whatever else grew along the road. The visit to Pompeii was very impressive. Back up through Naples and Rome again, passing Pisa, Monaco, Marseilles, Lyon, Paris, Reims, Belgium and finally, after 7 days, back in Eindhoven. He was barely back in time to get his four days of flying in and collect his pay for August. A lot skinnier, but proud of all the Italian he had learned while hiking around.

In Early September, another customer of Mama's pub caught Frits cleaning the place on Monday morning. Polishing all the chrome and all the glasswork was the least he could do for his room and board.

"Have you ever been to a "*Weinfest*" Frits?"

"I've never even been to a *Bierfest!*"

"Wanna go?"

"Wanna go where?"

"Do you know where Ahrweiler is?"

"Never heard of it!"

"Have you ever heard of the river AHR? It runs parallel to the Mosel and the other wine rivers in Germany. I believe it ditches into the Rhine at Cologne somewhere. Anyway, They are having their annual three-day *Weinfest*. It's a lot like "kermis" here and it started yesterday. We can be there by dark and come back tomorrow. Man, the wine just flows at those fests and the whole town goes crazy. It's a blast, you oughta see it!"

"Let me finish this bar a minute."

"See ya, Mam!"

On the way down in his old V.W. Dick asked: " Do you know any gals we can pick up and take with us?"

"Whassa matter? Aren't there any broads in Germany?"

"On Monday night? I don't know. It's very much like our "kermis" here. On Sunday and Tuesday, but mostly Sundays, you can't get near a bar for the crowds. Big cities, like Aachen and Cologne just stream into these little wine towns, but on Mondays it's mostly the locals, just like in St. Pancras, so there's a very slim chance of meeting some chicks. So who do you know?"

"Who do I know? Who's available on Monday afternoon? Most girls that I know work for a living. Are we coming through Arnhem or Nijmegen?"

"Arnhem, yes, but I'd like to avoid Nijmegen, but we can."

First stop in Arnhem... .BINGO!!!! Hannie, an attractive dark haired girl was home and willing and able. Several phone calls and visits had no results whatsoever. All her girlfriends were working, so they settled for Hannie.

She was an Indonesian-Dutch girl with some Chinese blood mixed in and a hell of a spirit.

She was a sex-demon if there ever was one!

They happily drank beer as the miles crunched away underneath the skinny Bug tires and the talk got dirtier and dirtier by the minute. Between the dirty jokes they told and the wild adventures Frits and Hannie relived, the whole gang got so horny that they could scream!

Result: Dick pulled off at some restaurant and the other two had to knock off a piece in the backseat of the little Volkswagen before they could even think of eating.

It had gotten dark by then and they were getting anxious,

tired and thirsty. The beer supply had run out long ago.

Finally at 9 o'clock they parked in the center of Ahrweiler, a charming little medieval town.

The first bar they saw was the first bar they hit.

"Ein Bier, bitte!" Frits was so thirsty, he thought he could drink a keg.

"'sGibts kein Bier am Weinfest, nur Wein!"

"Wein denn, bitte!"

They served the chilled new wine in big bells and Frits nearly downed the first one in one swig.

"Hey, this stuff is good!" "Ein anderes, bitte!"

The place was indeed reminiscent of St. Pancras at "kermis" time. No table cloths, old chairs, raw tables and boarded up windows. Long tables lined with people, swaying arm in arm, singing their hearts out! All the old waltzes and polkas that Frits grew up with even before the war, sounded out of the record machine, loud, but melodious!

They found a table and drank and danced and danced and drank! Dick was not the dancing type, he said, so he scanned the room for unattached females.

"Don't they never play a jitterbug?" Hannie wanted to know.

"Not in this neck of the woods. The fastest we're gonna get is a polka or a foxtrot, that's it!"

Wrong!

Of all things, they played the popular song from the Italian movie "Bitter Rice" with Sylvana Mangano: *"Tengo Ganar De Baillar El Nuevo Compas! Dicen Todos Cuando Me Ven Pasar, Chica, Donde Va?"*

Nobody got up to dance to that Latin beat, so Frits and Hannie had the floor ro themselves and danced their hearts out. The crowd loved it! They might never have seen that kinda dancing before and they shouted :"Bis, Bis" and applauded with all their hearts. "Noch mal, noch mal, noch einmal!"

Well, that really turned on the dancing pair and they did it "noch mal".

When they sat down, the questions rained down on them: "What do you call that dance? Where are you from? Where did you learn to dance like that?"

They promptly were invited to join a bunch of others at their table and they promptly joined in with the singing and swaying. A few waltzes with some of the older " Deutsche Frauen" and singing with the "Deutsche Männer" really got them in with the crowd. It turned out to be really good company, surrounded by some vineyard owners and even "Herr Bürgermeister" with his Frau. It made for a jolly group and they sang their hearts out.

At closing time the question was tossed around: "Do you people have a place to stay?" and when the answer proved negative, Frits and his "Frau" went home with one couple and Dick went home with another.

The house they landed in was big and square with a huge wine cellar with huge vats. Dozens of them. Twelve to fourteen feet long and ten to twelve feet high! What a sight!

Their host had them sample from different vats and tried to explain the differences. To Frits they all had the same taste: "Good! Very good!"

Upstairs in the kitchen they enjoyed another glass and then it was time to turn in.

The hostess handed Frits a bucket to take to the bedroom with the words: "You'll need this bucket. You're gonna throw up!" (Nemen Sie den Eimer mit, Sie wirden brechen!"

"Me? I never throw up!"

"Take the bucket anyway!"

Even though he felt a 100%, he took the bucket and stripped naked.

He was in bed in no time while Hannie was doing things to her face in the mirror over the little sink

While he was waiting for her to jump in, the ceiling started to turn and turn faster until he fell out of bed and threw his guts, his heart and his brains out in the bucket.

After emptying the bucket in the hall toilet and brushing his teeth he felt remarkably well and regretted not bringing some wine to bed. Hanny felt no pain and was as horny as ever.

In the morning, over breakfast, Frits was asked: "What do you do in Holland?" and when he told them, both of the hosts exclaimed: "Then you must meet Rudi!"

Rudi was the lady's brother and during the war he had been a "Luftwaffe" pilot who had managed to survive the entire war and Frits just had to meet him. Dick, who had joined them would rather go see the town than listen to "bullshit war stories" took Hannie with him to see the sights.

Well. Frits met with Rudi who seemed remarkably young.

Forty maybe. He asked endless questions about the "Düsenjäger" that Frits had flown, it's fire power, speed, rate of climb and maneuverability.

In return, Frits asked about Rudi's experiences and man, was he in for a treat! Rudi had hundreds of photographs, clippings, trophies and ribbons. He had shot down over a hundred aircraft, mostly Polish and Russian and had been shot down four times himself. Always over German held territory or he had limped back to the line, so he never got captured. It was surprising to learn how intense the air battles had been over Germany's Eastern front and how potent both the German and Russian Air Forces had been. He was finally captured by the Americans near the Austrian border and man, would he like to fly again some day.

What an interesting experience!

At lunch, Dick showed up without Hannie.

"Where's Hannie?"

"She had to get back, she took a train."

"A train? She didn't have any money, I thought! Did you give her money?"

Even though the conversation took place in Dutch, the Germans got the drift that something was wrong and with some hasty: "Vielen Dank!" and "Auf Wiedersehen", they were gone.

"No point in racing to the railroad station, she's gone!"

"I don't understand. What was her hurry all of a sudden?"

As they cruised back to Holland, the truth came out. He had taken her out of town and put the make on her, but she fought him off and he damn near raped her! She got out of the car and fled!

"You bastard! Where'd she go?"

"I don't know. She ran up on some path to a house and I scrammed!"

"You lousy bastard. By now that's three hours ago and we'll never find her. You'd better turn to Venlo up ahead and drop me off in Eindhoven. She might be hitchhiking home somewhere. You lousy rotten bastard!"

"She'll be alright!"

"Yeah, she's tough, but that doesn't keep you from being a lousy bastard!"

"But you were getting laid all the time and I didn't even… "

"Just shut up!"

CHAPTER XXVIII
THE FINAL FLIGHT

The final flight schedule called for Thursday, Friday, Saturday and Monday— and that was it.

He had his early discharge from his Reserve Contract and was ready to go.

Weather stunk and Frits decided to have his teeth checked at government expense, especially since he didn't know when he'd be able to afford a dentist in America.

2pm in the dentist chair on Friday: "Your teeth are fine, except, your one wisdom tooth has no opposite. You only have three wisdom teeth, so that tooth will just sit there and rot. On the other side, the bottom tooth is out of line, because there's no room for it and consequently the upper one has no function either and all three should come out!"

"OK!"

"OK what?"

"OK, take 'em out!"

"Fine, we'll do one today and Miss, schedule one for… "

"Wait Doctor ! I won't be here. I'm leaving for the States on the 27th. Take 'em all out now!"

"Do you know what you're in for?"

"Yeah I know. I won't be able to kiss any one tonight!" He winked at the pretty assistant.

"Hahahahaha, you won't kiss anyone for a week. Maybe even two, but let's do it!"

Two of the teeth came out reasonably easy, although it felt like the dentist was trying to break his jaw.

The third one was a bust. The crown broke off and the dentist had to cut into the gum and the bone in order to get to the root. For forty minutes, he hacked, picked and sawed in Frits' mouth, making him feel like his mouth was an anvil and a blacksmith was hammering horseshoes in shape. He was soooo dizzy when it was over that he nearly fell flat on his face when he got out of the chair, but with the help of the assistant he regained his composure.

"My God, she taller than I am.," he thought as he was leaning against her chest.

"No kissing tonight!" he said with a wry smile.

"OK, I'll have to wait till next week then!" She wasn't only tall, she had a sense of humor.

The doctor gave him some antibiotics and some pain pills.

"At about seven, the anaesthesia will have worn off , so at 6:30 start taking one of each and do that every three hours. You may still have a lot of pain, so see me first thing Monday morning. I'll want to take a look at it, OK? Have a good weekend!"

"I thank you. I think!"

At four o'clock he was sitting at the bar with Frans.

He had been in that dentist chair for nearly two hours. He could barely open his mouth and Frans couldn't sell him beer till 5.

"Do you wanna buy a bottle of jenever? Then it's yours and you can drink from it if you want to."

"That sounds good!"

When his buddies started drifting in from the flight line, they found him sipping Jenever from a little glass through barely opened lips.

They poked fun at his misery and he joked right along with them. When at 5:30 they all went into the mess, he stayed at the bar because he couldn't open his mouth wide enough to even put a spoon in.

Apparently, some of the guys must have talked to the chef, because of all things, he came out with a container of soup and a cup to drink it with.

To Frans he said: "Make sure he drinks all of his soup before you give him another drink, you hear?"

"Everybody is a goddamn doctor now, but chef, I got my own bottle of booze here!"

"Drink the soup, stupid!"

The guys filed back into the bar and even though it was Friday night, hardly anybody went into town, because of the stinking weather.

As usual, the singing and laughter started increasing with the amount of booze consumed and by 9 o'clock, his whole liter of Jenever was finished. (He did share some.)

Frits was back to beer and when the jukebox started belting out Harry Belafonte's "DEO", Frits played bongo's on a barstool and sang his heart out as if the operation had never happened.

Later on, he vaguely remembered two guys carrying him to his bunk, but he had no concept of time whatsoever.

A bursting bladder got him up in the morning and he waddled to the bathroom where the mirror reminded him of his missing teeth. His face was severely swollen on one side, as if he was hiding a tennis ball in his cheek, but he felt no pain! Only when he opened his mouth or touched his cheek did it hurt, but otherwise, no pain!

"Not bad!" he said to himself. "See what a little booze can do for a man?"

The barrack was empty. Everyone must have gone to the mess or the flight line already, so he gingerly shaved his face, showered and hit the mess. Most everyone was still there, sipping coffee, because there wouldn't be any flying today. The wind was whipping at 20 miles per hour and gusting to 45. Sheets of rain came slashing by and the clouds were racing through the tree tops. It stunk out there. It had been like that now for ten days. All over Europe, this air mass was dumping rain, rain and more rain. If this were November or March, it wouldn't be too unusual, but September? Strange weather patterns for this time of year! Holland with all it's rivers and canals and an excellent drainage systems, had very few problems. They simply pumped the water into the sea. But the rest of Europe was becoming a big morass. Along with that the A-Flu was raging westward from Russia, through Poland and Germany and now started invading Holland and Belgium. The weather wasn't helping any either.

The guys just leisurely hung around, waiting for the official word that they had been accounted for, for the day and could go home. Frits couldn't chew, but got a lot of French Toast and milk sucked into his stomach and was feeling remarkably well. He decided to hang around and maybe go visit his aunt and cousins and that way pass the time away.

At ten, a call came through the intercom in the mess: "We need a volunteer to fly!"

"Are they crazy? In this weather?" "They're nuts, I'm going home!" "Fuck them!"

Every remark was negative, except Frits'. "What's the mission?"

Hell, he looked ridiculous, his mouth throbbed a little and this might take his mind of it.

"Report to OPS!"

The mission was as follows: "With special permission from Air Force Headquarters, one light aircraft is to proceed to Güttersloh for customs check and onto a soccer field east-south-east of Güttersloh. Mission: Deliver penicillin and a diesel mechanic to the 4th Armored Division.

"You wanna do it Forrer?"

"Yeah, I'll do it. I look like shit anyway!"

"You feel up to it?"

"I feel fine as long as I don't move my jaw."

An Army mechanic was delivered and a case of vials with penicillin was strapped behind the passenger seat. At 11 o'clock, they were ready. With a mechanic "walking" each wing, they directed him into the wind and he was up in ten seconds flat.

The young mechanic, 19, maybe 20, told him he was from Nijmegen and he'd never been in a plane before, so Frits asked him" You wanna say hello to your Mam? Where do you live?"

The kid was game and the only concern Frits really had was high obstacles, like churches and power lines, etc.

He had highlighted them on the chart, but he hadn't planned to detour to Nijmegen, so they had to keep their eyes open.

It turned out to be a stupid move.

Finding the home of his mother was no problem. She lived right off the river not far from the famous bridge. Circling at

200 feet without hitting any churches was no problem either, but bouncing up and down in the back seat sure as hell became a problem for the young mechanic in the back. He threw his guts out. There was nothing that could be done about that, the mission had to go on.

Even though Frits had both hands on the stick, the plane bounced and danced all over the sky. Trying to keep it straight was an unbelievable and losing battle. They'd pitch and dropped ten to twenty feet at the time and the rain just came at them in sheets. Half the time he was on instruments at 500 feet.

Thank God for railroad tracks. He just stayed over the track leading east and only once did he startle when a church steeple flashed by, just 100 yards off his left wingtip.

According to his map, after 35 minutes, it was time to leave the tracks and turn north toward Güttersloh, a British CANBERRA base. In the old days the 314 squadron had been there on joint maneuvers, so the layout of the base was somewhat familiar.

Within 6 minutes they were over the base and he buzzed the tower to let the Tommies know they were there.

The wind was exactly perpendicular to the big runway, so Frits decided to land on it sideways. Even if there were no headwind, he could have made that, but with the wind howling straight at them, he damn near hovered over the runway like a helicopter.

British ground troops ran out and grabbed his struts and held him down on the tarmac and kept him from getting blown over before he was tied down. They covered the flyers with a poncho and rushed them into the hangar and out of the weather.

While waiting for custom officers to arrive and do their stupid thing, he tried to get some coffee, but all the lousy Limies had was lousy tea, so he bummed a ride to the club, hoping they'd have coffee on hand. Negative! Just more lousy tea!

"How about a rum and coke?"

"Not till after four, chap!"

Lousy Brits!

This seemed like the most idiotic thing he had ever been involved in.

There were patients waiting for penicillin, there was a hospital tent without power, because their diesel generator was busted and here they had to sit on an English base, waiting for some German custom people to arrive, to check if he wasn't importing something illegal. What could a little military plane possibly transport that was of such great national importance?

Stupid! Plain Stupid.

Worst of all, he drank some sloppy English tea with cream. Yech! It reminded him of the liberation of his home town in 1945. The Tommies and the Canadians liberated his part of the country and brought their lousy tea. His mother loved it, but without the cream.

Stupid, Plain stupid! That's all he could think of.

He made it back to the hangar and admired the Canberras, the high-flying British bombers that had embarrassed him many a time when flying an F-84 to 45,000 feet and watch the Canberras pass over them with at least a 5,000-foot clearance.

It was a beautiful plane though. While admiring the cockpit, the custom clowns finally arrived, looked through the Piper and cleared him for take-off. Wasted 21/2 hours.

He'd better hurry if he wanted to get back before dark.

The ground personnel walked him into the wind again and when he had full power, they turned him loose and he rose like a helicopter.

Thank God, the soccer field was right along a railroad track and fortunately, right near a rail road station where the name on the side told him he was at the right field.

The Army guys were pretty good. They had laid out a runway for him, straight into the wind with some potato and orange crates. The wind blew diagonally across the field, which gave him the greatest length, but the tallest trees were exactly where he'd be coming in.

A low pass over the field showed him all the details and warned the Army pukes that he was there.

The rain practically cut off all forward visibility, but by skidding a little bit, he could make out the field through the side window. The poor bastard in the back seat got sick all over again and he stunk to high heaven, but there was no time to worry about that.

Coming low over the trees in a skid, Frits cut the power and slid down to the ground, kicking the plane straight and losing all forward visibility at the same time. He hollered at the mechanic:

"If you have a God, this is the time to talk to him!" and pulled the stick in his stomach.

At that very moment, Army guys grabbed his struts and he was down on the ground safely."

"Halleluia! We pulled off another one!"

Thank God, the Dutch Army had coffee and they lent him some boots, so he could go admire the hospital tent.

What a sight! A tent, about 100 by 30 feet was lined with cots on both sides and there were the soldiers, sicker than dogs, suffering away as the water just ran into and out of the tent. The doctors and nurses, waded from bed to bed to do their things and it was a wonder that they weren't catching the worst pneumonia, working in these conditions. They worked by kerosene lamps, cause candles would have blown out inside the tent, it was so drafty.

They were glad to hear that a mechanic had been brought in and now Frits finally got to understand the importance of his mission. The roads were flooded, tanks were stuck in the mud, no ground transportation was possible and they didn't know how it would end or when.

The Army had their annual maneuvers going on with the English against the French and the Americans, but the only one winning the war was the FLU!. The whole kit and caboodle was bogged down in the mud. What a mess! It was a good thing that their food supply was right on hand, otherwise they might have starved as well.

Well, Frits was glad he could just fly away from all that suffering and be back in a cozy bar within hours.

Well… .that's what he thought! He wasn't going anywhere!

"What the hell is this? I can't fly out? Why the hell not?"

"I have no orders? Yes I have orders. How else do you think I got here?"

"Yes, Sergeant, you had orders to fly here! Not to fly back!"

"Who the hell dreamed that up?"

"There's no point in getting exited. Air Force Headquarters was very specific. You stay here till the weather clears!"

"Till the weather clears? That may be in November and I'm going to be on a ship on September 27, so I'm leaving!"

"I'm afraid not." The Colonel was very patient and soft spoken, but also adamant. "We have arranged quarters for you with our N.C.O.'s and let's hope that by Monday, the weather has cleared enough so you may return. Meanwhile, we greatly appreciate your accomplishments and if it were up to me, I'd recommend you for a medal, but for now you are our guest. Dismissed!"

"What do you want me to do, hang around here till I also catch that damn flu."

"Dismissed!"

What a rotten deal! To the Sergeant-Major in charge; "If they had told me that I couldn't return, I would not have taken on this mission. Those lousy bastards! Can't they never tell the truth?"

He was mad as a pistol, but there was nothing he could do. He couldn't just walk out there, untie the plane and take off. The plane was well guarded and protected by the few Army guys that were still walking.

The N.C.O. tent was more solid and more cozy then the hospital tent and within an hour, the power came back on. The tent was also on higher ground at least and his fellow campers were at least friendly and sympathetic. One of them even said that his "troop carrier" could make it into town and that he needed to go for supplies if they could find any and maybe a game of billiards and some beer.

"Now you're talking, man!"

The "town" was not even a village. It consisted of maybe 50 houses, that included a store and "thank Heaven", a little pub with a pool table. Because of the weather, there was not a soul in the place, except the owner, who tapped the beer. According to the owner, "Herr Wirt", any sizable town with some action was not within range of a "troop carrier" and the roads were just about all impassible.

What a mess!

They played cards, told jokes and sipped on the beers they'd brought, even though that was against regulation. "Just blame it on the Air Force!"

The next day was much of the same, although the rain let up a little and they did make a church-bar trip into "town" and a return trip in the evening.

On Monday, rain was intermittent, the clouds were all the way up to 800 feet and Headquarters OK'd his return flight. This time it was much simpler. Again to Güttersloh for custom check, "Stupid, how stupid!", and then back along the railroad track but at least at 500 feet.

It was still bumpy and the wind gusts would try to throw him all over the sky, but compared to Saturday, this was a picnic. His face was still swollen, but he experienced practically no pain.

Obviously, he missed his early morning appointment with the dentist, but they were glad to see him at about three. The dentist was pleased and surprised, that he hadn't experienced more pain.

"You see, Doc, you should prescribe Jenever instead of antibiotics!"

"I just may, I just may."

The tall assistant with the long hair was also very amused and that encouraged him to suggest a drink after work. Of all things, she said yes! She lived in Nijmegen and shuttled back and forth by train, so they agreed to meet in a pub across from the railroad station at about 5:30.

It was the same little pub where he had met with Marion eons ago and the memory sure got his imagination going. This gal was beautiful and maybe 24,or 25 years old.

Apparently, she had nobody permanent in her life which seemed very surprising. As beautiful as she was and in constant contact with young military men, you would think that she had dates coming out of her ears.

In spite of the fact that he had barely ten days left in Holland, they agreed to meet again shortly, but this time in Nijmegen, where she lived with her folks.

As he rode the train back to St. Pancras, he kept wondering how come he had never asked her before. He had seen her often enough on the base, but she seemed somewhat aloof and stuck-up, but that was not the case at all. She was as down to earth and pleasant as one could wish for.

There was a certain amount of running around to do. The sponsor papers from Pieter Groot in Texas had come in and needed to get to The Hague, Catholic Charities in Amsterdam had his boat ticket and gave him $35 landing fee. (Thank you very much!)

All he packed for America was his old B-4 bag, that survived all his travels and he needed to muster out at Deelen and turn in his uniform and other military belongings.

Just one drink in the bar, just fleeting good-byes and he was on his way to Nijmegen and his newfound love.

Her folks were most pleasant and regretted that he was going to leave so soon.

They danced the night away, smooched in the park and walked back to her house early enough for him to take the last train out. They made plans to get together in Rotterdam the following week and then she would send him off the next morning.

Man, that sounded promising!

The days flew by, but when he called her for details as to where and when to meet, she made excuses and he hung up, disgusted.

Inasmuch as he had convinced his mother that he had to be in Rotterdam the night before embarking, he said goodbye to everyone at home and got on the bus.

No point in going to Rotterdam alone, so he continued on to Eindhoven, latched on to Beppie and spent the evening and the night with her for old times sake.

Hanging over the railing, watching the dock fade away, he had very mixed emotions.

Happy and excited about going to his new homeland, but melancholy and sad about leaving all his friends, family and great times behind.

He sang to himself: "It Was Great Fun, But It Was Just One Of Those Things!"

Indeed, it had been exciting. All the girls, all the parties, but most of all:

THE FUN OF FLYING!

The End

About the Author

After living with near-daily bombings during World War II, Frits decided to become a Fighter-Bomber- Pilot himself, earning his Silver Wings with the United States Air Force in October of '53.

Still an active Pilot, he has lived in the United States continuously since '57, bringing over four of his brothers and sisters. All are married and now have a total of 43 children (including in-laws) and grandchildren. A true pioneering family. He now writes and boats in the quiet community of Gulf Breeze in the Florida Panhandle.

CPSIA information can be obtained at www.ICGtesting.com
Printed in the USA
LVOW12s0719140414

381538LV00001B/1/P

9 780971 449039